LOW-LEVEL RADIOACTIVE WASTE REGULATION

Science, Politics and Fear

LEWIS PUBLISHERS, INC.

LOW-LEVEL RADIOACTIVE WASTE REGULATION

Science, Politics and Fear

Edited by

Michael E. Burns

LEWIS PUBLISHERS, INC.

Library of Congress Cataloging-in-Publication Data

Low-level radioactive waste regulation: science, politics,
and fear / Michael E. Burns, editor
    p.   cm.
    Bibliography: p.
    Includes index.
    ISBN 0-87371-026-6
    1. Radioactive waste disposal—United States. 2. Radioac-
tive waste disposal—Government policy—United States.
I. Burns, Michael E. (Michael Edward), 1952–
TD898.L676   1988
363.7'28—dc19   87-30681

Third Printing 1989

Second Printing 1988

LEWIS PUBLISHERS, INC.
121 South Main Street, Chelsea, Michigan 48118

PRINTED IN THE UNITED STATES OF AMERICA

This book is dedicated to my father for teaching me to think and encouraging me always to ask why.

Contents

Preface

Although radioactivity is a natural phenomenon, scientists did not discover its existence until 1896. In the ensuing 92 years, radiation and radioactive materials have come to play an integral part in our everyday lives. But few people seem to understand or even give much thought to the importance of the role that they play.

An inevitable consequence of the use of these materials is the generation of radioactive wastes and the public policy debate over how we will manage them. In 1980, Congress shifted responsibility for the disposal of low-level radioactive wastes from the federal government to the states. This act represented a sharp departure from more than 30 years of virtually absolute federal control over radioactive materials. Though this plan had the enthusiastic support of the states in 1980, it now appears to have been at best a chimera.

Radioactive waste management has become an increasingly complicated and controversial issue for our society in recent years. This book discusses only low-level wastes, however, because Congress decided for political reasons to treat them differently than high-level wastes. We hope this volume will clarify some of the complex social, political, and scientific conflicts that have resulted from this change in policy.

This book is based in part on three symposia sponsored by the Division of Chemistry and the Law of the American Chemical Society. Each chapter is derived in full or in part from presentations made at these meetings. The book is not, however, simply a recapitulation of these symposia. From the outset, we decided to write it for an audience of both scientists and nonscientists. Our intention was to provide a reasonably comprehensive review of the subject in nontechnical language and a readable style. The readers will be the judges of whether we have succeeded.

Acknowledgments

Many people are owed my gratitude for their assistance in the course of this project. It is impossible for me to name each individual who contributed directly to its completion or provided me with moral support and encouragement. To all of you who were a part of this project, I extend my sincere appreciation for your efforts.

I must first thank my colleagues on the Executive Committee of the American Chemical Society's Division of Chemistry and the Law for providing me with financial support and encouragement throughout this endeavor. Without their help and continued moral support, I could never have completed this book. Although I am reluctant to single out individual members of a group that has provided me with so much support, I must mention the contributions of Howard Peters and Rose Ann Dabek. They unselfishly provided me encouragement when I needed it most. Most of all, they were there when I needed them. Both deserve and have my deepest appreciation.

Next I must thank my former colleagues in the Environmental Control and Research Program of the National Cancer Institute's Frederick Cancer Research Facility. From the beginning of this project, they gave me support and encouragement, and perhaps most important, they patiently listened as I bounced ideas off them. Many of them now know more about this subject than they ever wanted to know. Out of this group, Ev Hanel and Bruce Tobias were my most patient (or perhaps tolerant) listeners.

Equally important has been the support of my present colleagues in the Research and Test Division of the Association of American Railroads. Once again, they have probably heard too much about this subject — but they continue to listen patiently.

I owe a special debt to Ethel Armstrong of the NCI-FCRF library who was instrumental in locating and obtaining innumerable obscure references and did so cheerfully and efficiently.

Hannah King, of the Department of Energy's library in Germantown, Maryland, also provided valuable assistance in locating key personnel and references.

Naturally, I am also indebted to each of the chapter authors for their work in putting together both their presentations and the written material. Many were helpful behind the scenes in providing technical information, recommending other contributors, or using their contacts to smooth my path. My thanks to you all.

I must also thank Ed Lewis for his initial and continuing faith in this project and Jon Lewis, my editor, for his guidance and assistance.

My special thanks are due Michalann Harthill, who has been my colleague and friend for several years. Michalann was the editor of my first publication. Without her pushing early on, I might never have written that first chapter.

Last, but decidedly not least, I would like to thank my wife, Jane Banks, for cheerfully putting up with me and this project. Jane has been my greatest source of support and encouragement throughout this task. She too has heard more about this subject than any individual should have to bear. In addition to acting as a sounding board, she has been my reference librarian (usually on short notice with the information needed yesterday), reviewed and edited endless chapter drafts, and contributed in many other ways. In addition, though my name appears alone as editor of this volume, a project of this magnitude is a family project, and Jane has eaten alone many nights and done more than her fair share of the household chores while I worked on this book. This is as much her effort as mine.

Having completed this project, I look back and realize that without the help of the people I have mentioned and many others that I have not, this book would never have been completed. At least not by me. Now that it is complete, I have realized that it has been one of the most rewarding experiences of my life. That fact alone should tell all of you who have helped me just how much I appreciate your efforts.

About the Editor

Michael E. Burns is employed as an environmental chemist by the Research and Test Division of the Association of American Railroads. He was formerly the chemical safety officer for the National Cancer Institute's Frederick Cancer Research Facility and safety and regulatory compliance officer for a hazardous waste management facility.

For the past two years, he has maintained a private consulting practice in the areas of chemical safety and emergency response and has served as a training consultant to the J. T. Baker Chemical Company.

Mr. Burns holds a BS in chemistry (1980) from the University of Maryland and an MA in environmental biology (1985) from Hood College.

He has been an active member of the American Chemical Society's Division of Chemistry and the Law, serving on the division's executive committee and as its representative on the editorial advisory board of the ACS journal *CHEMTECH*. He is a member of the American Society of Safety Engineers, the National Fire Protection Association, and the Association of Ground Water Scientists and Engineers.

His professional interests include the development of integrated chemical safety programs, public education about scientific issues, and the interface of chemistry and the law, especially in the development of regulatory policy.

About the Authors

Richard J. Bord is an associate professor of social psychology in the Department of Sociology at The Pennsylvania State University. Since 1983, his work has focused on public involvement in low-level radioactive waste disposal siting. He has published articles on public involvement in technical journals, waste management journals, and a number of edited volumes. He was the principal investigator in a statewide study examining attitudes toward waste policy issues among the general public and influential decisionmakers. This study has been made available to all state legislators and has been cited both nationally and internationally. Papers on public involvement have been presented to the American Nuclear Society, the American Sociological Association, the American Chemical Society, Waste Management/86, the American Association for the Advancement of Science, and the American Petroleum Institute. In addition to the above, Professor Bord did research in communities facing a waste treatment facility and in communities facing the startup of a nuclear power facility. He has had the opportunity to see citizen protest groups mobilize and to discuss issues with their leadership. This experience was put to good use in traveling throughout Pennsylvania as a member of the Public Information and Education on Radiation (PIER) program for the last three years.

William H. Briner is an associate professor of radiology and director of the Duke University Medical Center Nuclear Medicine Laboratory and Radiopharmacy. He is chairman of the Society of Nuclear Medicine's Government Relations Committee and has served as a commissioner (North Carolina) and secretary-treasurer of the Southeast Compact Commission for the last several years.

E. William Colglazier, a physicist, is director of the Energy, Environment, and Resources Center at the University of Tennessee, Knoxville. He also directs the Waste Management Research and Education Institute, a new state-sponsored "center of excellence" at the university. He has edited a book, *Politics of Nuclear Waste*; chaired a task force for a presidential commission on radioactive waste management; and is a member of the Board on Radioactive Waste Management of the National Academy of Sciences. He received his PhD in physics from Caltech and was a researcher at the Stanford Linear Accelerator Center and the Institute for Advanced Study (Princeton). He served as associate director of the Aspen Institute Program in Science, Technology, and Humanism, and spent five years at the Kennedy School of Government at Harvard working on science, technology, and public policy research.

John O. Eichling is an associate professor of oral diagnosis at the Washington University School of Dental Medicine in St. Louis, Missouri. He has been associated with the university since 1963. Prior to that, he was an assistant professor of physics at Northeastern Oklahoma State College.

He holds a PhD from Washington University, an MS from Oklahoma University, and a BS from Northeastern Oklahoma State College.

Mary R. English is an assistant director of the Energy, Environment, and Resources Center. She has worked for state government in environmental planning and has contributed to publications in the fields of land management and environmental policy. She currently is part of a five-member research team headed by Dr. E. William Colglazier that is analyzing values issues in radioactive waste management under a grant from the National Science Foundation. She holds a BA in English from Brown University and an MS in regional planning from the University of Massachusetts.

Lawrence M. Gibbs is director of the Department of Chemical Health and Safety at Yale University as well as an adjunct faculty member at the Yale University School of Medicine, where he teaches courses in industrial hygiene and environmental health. He was previously an environmental health specialist at the University of Connecticut Health Sciences Center, a private consultant in the area of industrial hygiene, and an industrial hygienist for the U.S. Occupational Safety and Health Administration.

Mr. Gibbs holds a master of public health degree from the University of Michigan's School of Public Health as well as MEd and BA degrees.

He is a certified industrial hygienist and certified hazard control manager. He is a member of numerous technical and scientific societies as well as the author of fifteen publications in the area of occupational and environmental health.

Robert L. Glicksman is a professor of law at the University of Kansas School of Law. He received his AB from Union College in Schenectady, New York, in 1973; his MA in history from Harvard University in 1974; and his JD from Cornell Law School in 1977. After four years in private practice with a Washington, DC, law firm, Professor Glicksman taught as a visitor at the University of Arkansas School of Law in Fayetteville for one year and has taught at Kansas since 1982.

Professor Glicksman has written extensively in the area of environmental law. He coauthored a book on strategies to prevent groundwater pollution in 1986 and a book chapter on municipal regulation of hazardous waste in 1987. He has published articles on federal public land management, private legal remedies for pollution, groundwater pollution, and the development of environmental law in the federal courts in such journals as the law reviews of the University of Pennsylvania and University of Kansas; *Chicago-Kent Law Review; Hastings Law Journal;* and *Columbia Journal of Environmental Law.*

George R. Holeman is director of the Health Physics Division of Yale University. He has held this position since 1971; he has been affiliated with Yale since 1963.

Mr. Holeman holds a BA degree in physics/mathematics from the Centre College of Kentucky and an MA in engineering/health physics from Harvard University.

He is a member of numerous professional societies.

Letty G. Lutzker, MD, is a board certified radiologist and nuclear physician who is currently associate director of radiology and chief of nuclear medicine at Woodhull Hospital in Brooklyn, New York. She is an active member of the Society of Nuclear Medicine and has represented the Greater New York Chapter of that society on the New York State Low-Level Waste Group for the past five years.

Dr. Lutzker received her MD from the Albert Einstein College of Medicine and a BA from Smith College. She holds a number of academic appointments in the areas of radiology and nuclear medicine, is a member of a number of professional societies, and is the author or coauthor of numerous publications.

Carla J. Mathias is a research assistant in the Division of Radiation Sciences at the Washington University School of Medicine in St. Louis, Missouri. She holds a BA in zoology/chemistry from Depauw University.

John M. Matuszek is director of the Radiological Sciences Laboratory of the New York State Department of Health. He is also an adjunct professor of Nuclear Engineering at Rensselaer Polytechnic Institute, where he teaches a course on radioactive waste management.

He is a member of the National Academy of Sciences' Board on Radioactive Waste Management, Panel for the Study of Waste Management Practices at Oak Ridge National Laboratory. He is also chairman of the Task Force on Low-Level Radioactive Waste for the National Council on Radiation Protection and Measurements.

Dr. Matuszek's degrees include a BS in chemistry (with distinction) from Worcester Polytechnic Institute and a PhD in nuclear chemistry from Clark University.

Frank L. Parker is professor of Environmental and Water Resources Engineering and Management of Technology at Vanderbilt University. Prior to going to Vanderbilt in 1967, he was chief of the Radioactive Waste Disposal Research Section at the Oak Ridge National Laboratory. He was previously senior officer at the International Atomic Energy Agency in Vienna, and head of their Waste Disposal Research Section. At the present time, he is the chairman of the Board of Radioactive Waste Management of the U.S. National Academy of Sciences, senior research fellow of the Beijer Institute of the Royal Swedish Academy of Sciences, and director of the Pilot Internship Program in Decontamination and Decommissioning of Nuclear Facilities at Vanderbilt University.

Timothy L. Peckinpaugh is an associate with the law firm of Preston, Thorgrimson, Ellis & Holman of Washington, DC, working on energy legislative issues related to the congressional budget process. His particular expertise, on which he has published articles and given several presentations, is nuclear energy.

Mr. Peckinpaugh spent five years working on Capitol Hill as a technical consultant to the House Committee on Science and Technology and as a legislative assistant to Congressman Sid Morrison (R-Wash.).

He holds a BA degree from Claremont Mens' College and a JD degree from Georgetown University Law School.

Don G. Scroggin, a partner in the Washington, DC, law firm of Beveridge & Diamond, PC, holds a PhD in chemistry from Harvard University and was for four years a chemistry professor at Williams College and Harvard, before resigning to attend Yale Law School. He served two years on the staff of the White House Council on Environmental Quality in Washington before joining Beveridge & Diamond. He has been deeply involved in scientific risk assessment and risk management issues, particularly as they relate to environmental regulations. Besides serving on editorial boards, he has published numerous articles in scientific and legal publications regarding environmental law and, in particular, its use and misuse of science. In his environmental law practice, his firm represents parties interested in the use of health risk assessments in government regulations.

Robert Shaw is the senior program manager of the Low-Level Waste and Coolant Technology Program at the Electric Power Research Institute (EPRI). Dr. Shaw joined EPRI in 1975 as a member of the technical staff in the Nuclear Power Division. He assumed his present position in 1981.

Dr. Shaw was formerly a consultant to the Nuclear Energy Division of the General Electric Company as well as a professor of chemical engineering at Clarkson College of Technology in Potsdam, New York.

Dr. Shaw holds BS and MS degrees in chemical engineering and a PhD in nuclear science and engineering from Cornell University. He is a member of the American Nuclear Society and the American Institute of Chemical Engineers and is the author of numerous papers.

Barry A. Siegel, MD, is a professor of medicine at the Washington University School of Medicine in St. Louis, Missouri. He also holds a number of clinical and administrative appointments. He served as a major in the U.S. Air Force, holding the post of chief, Radiological Sciences Division, Department of Radiation Biology at the Armed Forces Radiobiology Research Institute in Bethesda, Maryland.

He received his MD from the Washington University School of Medicine in 1969. He is a member of numerous professional and scientific organizations.

Michael J. Welch is a professor of radiation chemistry at the Mallinckrodt Institute of Radiology at Washington University in St. Louis, Missouri. He has been associated with the university since 1967.

Dr. Welch holds a PhD from the University of London and MA and BA degrees from Cambridge University. He was president of the Society of Nuclear Medicine in 1984. He is a member of a number of professional and scientific organizations, including the National Academy of Sciences.

Rosalyn S. Yalow received her AB from Hunter College in 1941 and an MS (1942) and PhD (1945) from the University of Illinois in nuclear physics. She joined the Bronx Veterans Administration Hospital in 1947 and since 1972 has been a VA senior medical investigator. She is director of the Solomon A. Berson Research Laboratory at that medical center and is the Solomon A. Berson Distinguished Professor-At-Large of the Mt. Sinai School of Medicine of the City University of New York.

Dr. Yalow received the Nobel Prize in Physiology or Medicine in 1977 for the development of radioimmunoassay. She is a member of the National Academy of Sciences and the American Academy of Arts and Sciences. She has been the recipient of numerous honors and awards, including 39 honorary degrees from universities in the United States and abroad.

CHAPTER 1

Setting the Stage

Michael E. Burns and William H. Briner

Late on the afternoon of Friday, November 8, 1895, the rector of the University of Würzburg gladly put aside his administrative tasks and retired to his laboratory. He had always found his administrative duties boring, preferring instead to conduct laboratory research. On this cloudy afternoon, he spent several hours preparing his experimental apparatus by the light of the laboratory's gas lamps. The shy, 51-year-old physicist was only attempting to extend the experimental results of a colleague. Instead, he discovered a phenomenon that forever changed his life and the world we live in. Wilhelm Conrad Roentgen was about to become the most famous scientist in the world — as a result of serendipity.

This chapter reviews some of the events that led to the discovery of radioactivity, early radiation research, radioactive waste disposal policies, and the decisions that led to our current national policy on low-level radioactive waste (LLRW) management. This chapter does not pretend to be comprehensive; each of its subject areas have been reviewed thoroughly in other publications. The intention is to provide a broad overview of the subject to enhance the other chapters in the book.

Nineteenth-century scientists were intrigued by the properties displayed by electricity, such as the ability to change matter as it moved through it. What caused these changes was the subject of much speculation and many experiments. These experiments, however, were often hampered by the unreliability of equipment until the middle of that century. In 1854, Heinrich Geissler, a German glassblower, fabricated a forerunner of the modern electron tube. (Electron tubes used to glow in radios and televisions.) Geissler's device consisted of an evacuated[1] glass tube with a metal electrode sealed in each end. When a voltage was placed across the electrodes, electrons flowed from the cathode (the

Low-Level Radioactive Waste Regulation: Science, Politics, and Fear, Michael E. Burns, Ed., © 1988 Lewis Publishers, Inc., Chelsea, Michigan — Printed in USA.

negatively charged electrode) to the anode (the positively charged electrode).[2] Geissler used his tube to investigate the effects of an electrical discharge through rare gases. With certain gases he saw a reddish glow around the anode and a violet glow around the cathode. When he improved the tube's vacuum, the red glow disappeared and the violet color became a beam.[3] Since the beam appeared to originate at the cathode, he called it a cathode ray. (This manuscript was typed on a computer with a video display terminal that is a direct descendant of Geissler's tube.) Later, other scientists, notably Hittorf and Crookes, designed variants of Geissler's tube.

In a follow-up to Geissler's work, Sir William Crookes, an English chemist and physicist,[4] demonstrated by 1885 that cathode rays were negatively charged, traveled in straight lines, and generated intense heat when they struck an obstacle.[2]

Because of the depth and breadth of his research with cathode ray tubes in the 1870s, Crookes's name is most commonly associated with the device.[5]

MYSTERIOUS RAYS

Wilhelm Conrad Roentgen [6] was the first person to discover and recognize the existence of a radioactive emission.[7] Born in Prussia in 1845, Roentgen received his PhD in physics from the University of Zurich in 1869.[8] He was a professor of physics and rector of the University of Würzburg when he accidentally discovered that rays capable of penetrating matter were emitted when cathode rays struck glass. Prior to this discovery, Philipp Lenard, a professor of physics at the University of Heidelberg, had established that cathode rays could pass through an aluminum foil window in a cathode ray tube and travel a few centimeters in air.[9]

Roentgen was attempting to extend Lenard's work on that November afternoon. Specifically, he wanted to determine if cathode rays were capable of passing through the thick glass wall of a Crookes tube.[10] Roentgen's experiment was not the first attempt to answer this question, and he was aware that other experimenters had not been successful. As it turned out, Roentgen's results did not settle the question either. Instead, he discovered something far more significant.

Roentgen's experimental procedure differed only slightly from that of others in that his laboratory was dark, the tube was covered with a sleeve of black cardboard, and a screen coated with barium platinocyanide happened to be on a nearby table. Before starting the experiment, Roentgen wanted to check the opacity of the tube's cover. He turned off the gas lamps and energized the tube to see if any light showed through the black cardboard cover. As soon as he energized the tube, a faint greenish-blue glow appeared in the corner of the room. The light disappeared when he turned the tube's power off. When he energized the tube again, the light reappeared. This time, he watched it carefully, and noted that the light's intensity varied with the pulsations of current to the tube. By the light of a match, Roentgen discovered that the glow was fluorescence from the screen coated with barium platinocyanide. He moved the screen farther away from the tube — yet the screen continued to glow, even when it was 6 to 7 feet from the tube. Beyond that point the fluorescence rapidly diminished.

Since Lenard had already established that cathode rays could travel only a few centimeters in air, Roentgen correctly concluded that a previously unknown type of emission must be causing the fluorescence. He decided to call the mysterious emissions X-rays ("X" for the unknown in algebra).

"I Have Discovered Something Interesting . . . "

Puzzled and excited by his results, Roentgen spent most of the next two months working alone in his laboratory. Because he couldn't explain his results, he refused to tell any of his colleagues the details of his discovery. To the amazement of the university community, he put a cot in his lab, had his meals delivered (often returning them untouched), and virtually lived in his laboratory for several weeks.[3] When a close friend asked him somewhat irritably what he was working on so secretly, Roentgen, in a classic piece of scientific understatement replied, "I have discovered something interesting, but I do not know yet whether or not my observations are correct."[8]

Roentgen found that though X-rays travelled in straight lines like cathode rays, they were not deflected by a magnet, and thus they were not electrically charged. He also discovered that the rays penetrated paper, wood, cardboard, and even a 1,000-page book placed between the tube and the screen. None of the objects appreciably weakened the screen's fluorescence. He tested other items readily available in the physics department and found that X-rays could penetrate most thin metal foils, but were blocked by various thicknesses of aluminum, platinum, and lead.[3]

While testing lead's ability to block the rays, Roentgen held a solid disk of lead in front of the screen. To his surprise, the screen showed not only the disk's shadow, but also the outline of his hand, with the bones appearing as distinctly darker shadows. He then tried substituting a photographic plate for the screen and discovered that the X-rays exposed the plate except where an object blocked the rays. Intrigued by the sight of the bones of his hand, he persuaded his wife to assist him. He put her hand on top of a film plate and exposed it to X-rays for 15 minutes. The developed plate clearly showed the bones of her hand as well as the two rings she was wearing. An X-ray picture of a woman's hand, probably Frau Roentgen's, made by Roentgen on December 22, 1895, exists today.[11]

Roentgen's Announcement

Roentgen announced his discovery as a "preliminary communication" in a letter to the Würzburg Physical-Medical Society, which published it on December 28, 1895. Because he had not discussed his findings with his peers, Roentgen was not sure how his discovery would be received. He sent reprints of his paper to a number of prominent European physicists, some of them friends, on New Year's Day of 1896, commenting to his wife, "Now the devil will be to pay."[8]

As it turned out, Roentgen need not have worried about the reaction of his peers. Almost overnight he became an object of worldwide attention and acclaim. Imagine this modest man, who subordinated himself to his science, suddenly finding himself the most famous scientist in the world. A veritable horde of reporters, scientists, and curious members of the public flocked to see Roentgen and his laboratory. Some visitors stole X-ray plates from his lab, and postcards bearing his signature that he sent to friends failed to arrive (presumably intercepted by souvenir hunters).[8]

Roentgen was deeply disturbed by these transgressions of his privacy, as well as the practice of the press and other scientists of referring to X-rays as "Roentgen's rays" and X-ray photographs as "Roentgenographs." The use of his name in conjunction with a natural phenomenon deeply disturbed him. He preferred to call the photographs shadowgrams or skiagrams (skia from the Greek word for shadow), and consistently used these terms for the rest of his life. Roentgen often stated that he felt that the popularization of science did far more harm than good, and that he could not understand why people were interested in him rather than his discovery.[8]

The discovery of X-rays rocked the scientific world. X-ray research programs were started all over the world, both to investigate the ray's characteristics, and to find uses for them.

Roentgen refused to patent his discovery and turned down all of the many offers he received to exploit it commercially. He was awarded the first Nobel Prize for Physics in 1901, and kindly donated the prize money to the University of Würzburg, to further scientific research.

In his later years, Roentgen became something of a recluse in the face of a series of bitter ad hominem attacks from a small group of fellow scientists. They claimed that others had set the stage for Roentgen's discovery and that he merely blundered into it. One apparently malicious story held that Roentgen's laboratory assistant *(diener)* was actually the one to discover the phenomenon, and that Roentgen had stolen the credit for it. Philipp Lenard was one prominent critic who claimed that Roentgen's discovery resulted from Lenard's work. Any motive for Lenard's attack other than jealousy is difficult to understand, since Roentgen always gave Lenard credit for the research on cathode rays that paved the way for the discovery of X-rays. These attacks from men Roentgen had considered both colleagues and friends left him bitter and withdrawn. He died on February 10, 1923, of rectal carcinoma. Regrettably, in a great loss to science, all of his notes and papers were burned after his death in accordance with his will.[8]

THE DISCOVERY OF RADIOACTIVITY

Antoine Henri Becquerel

When he heard of Roentgen's discovery, French physicist Antoine Becquerel wondered if the fluorescence was induced by ordinary light or by X-rays. He decided to study salts of the rare metal uranium because they were known to phosphoresce after exposure to light. Like Roentgen, Becquerel was lucky. By choosing uranium, rather than a nonradioactive phosphorescent compound, he became the first to observe and note the existence of natural radioactivity.

On January 20, 1896, Becquerel taped a piece of uranium bisulfate to the outside of a film cassette which he placed on his windowsill. The developed film clearly showed the outline of the wafer. He repeated the experiment several times, once placing a 10-mm-thick glass plate between the wafer and the cassette. Each time, the outline of the wafer was present on the exposed film. He published these results on February 4, 1896.

Since the cassette prevented any light from reaching the film, Becquerel assumed that the uranium salt emitted something that exposed the film. However, he still didn't know whether the film exposure was caused by fluorescence or by X-rays. Like Roentgen, he would make his greatest discovery by accident.

Becquerel continued his experiments during February 1896. On February 26th and 27th, Becquerel prepared several film cassettes with attached uranium wafers. However, these days were largely overcast with the sun only intermittently visible. He put the plates away in a dark drawer with the wafers still in place. On Sunday, March 1, he took them out to use, but, almost on a whim, he decided to develop the old ones and work with fresh cassettes. Since the cassettes had only been exposed to a few minutes of sunlight, he expected at best to find a weak impression of the wafers. Instead, he found the photographic impressions to be very clear—by far the most distinct that he had obtained to date. Now he was certain that the emission given off by the uranium salts had nothing to do with exposure to light, and quickly reported his results to the French Academy of Sciences.

Becquerel's discovery, coming on the heels of Roentgen's, intensified public and scientific interest in radiation. Many theories were advanced to explain the source of the newly discovered emissions. Some thought they resulted from a chemical reaction, others argued that they were somehow induced by light, and yet another school of thought felt that they were a change taking place at the atomic level. Marie and Pierre Curie were of the latter opinion.

The Curies

Pierre Curie was chief of the laboratory at the School of Physics and Chemistry of the City of Paris in 1895. His wife, Marie Curie-Sklodowska, was a doctoral candidate in physics at the University of Paris. In 1896, after the announcement of Becquerel's discovery, Marie Curie decided to study the fundamental properties of the rays given off by uranium as part of her PhD dissertation. Her initial hypothesis (which was not particularly popular with her advisers) was that the emission was a result of an atomic change—a function of a change in the atom itself, rather than a chemical reaction.

Marie immediately encountered problems finding laboratory space for her research. There didn't seem to be any space available for the school's first female PhD candidate, especially a married woman with a child, who was on what appeared to be wild goose chase. Finally, with Pierre's help, she was given permission to convert a small storeroom on the ground floor of his institution into a laboratory. The room was damp and unheated, and the roof leaked. It was available to her only because no one else wanted it. One February day, she recorded the room's temperature in her laboratory notebook as 6°C (43°F).[12]

At the beginning of her research, Marie discovered that the intensity of the uranium emissions was proportional to the amount of uranium present, regardless of which uranium salt she tested. This suggested that elemental uranium was the source of the rays. She also learned that the emission's intensity was not affected by external factors, such as room temperature or the amount of light present. Now, Marie felt even more strongly that the emanations were an atomic property. If this were the case, why should uranium be the only element to possess this property?

To test this new hypothesis, she examined all known elements looking for any with similar behavior. This was an arduous task that conceivably could have taken years. One might imagine how hard it must have been to start this work. After all, her decision was based only on a hypothesis, a scientific guess, that other elements might possess properties similar to uranium's. She soon found that thorium compounds also gave off spontaneous emissions. In the end, only uranium and thorium compounds yielded positive results.

Now, restricting her work to uranium and thorium compounds, Marie found that the intensity of the emissions was far stronger than might be presupposed by the amount of uranium or thorium present in the compounds. Suspecting that her experimental results were in error, she repeated her calculations many times and had Pierre check them. They found no fault in her work. If her calculations were correct, then the only tenable hypothesis to explain the anomaly was that there must be an unknown substance, present only in minute amounts in uranium and thorium compounds, that caused such a powerful emission.

The New Elements

Since Marie had already examined all the known elements, the unknown, if it existed, had to be an undiscovered element. This was a daring hypothesis since many of her contemporaries still thought she must have made an experimental error. Pierre demonstrated his confidence in her by choosing to put aside his own research on crystals and join Marie in the search for the new element. They were to remain collaborators for the next eight years, until Pierre's untimely death.[12]

Their uranium salts had been isolated from pitchblende ore obtained from Czechoslovakian mines. On April 12, 1898, Marie boldly suggested in a publication the probable existence of an undiscovered radioactive element in pitchblende ores.

The Curies' search for this new element used classical chemical methods to separate the pitchblende from the ore and into its individual elements. They then measured the amount of radioactivity present in each component. In July 1898, they established that the radioactivity was concentrated in not one, but two different chemical fractions of the ore.

The obvious conclusion was that there were two new elements instead of one.

When they isolated the first new element as a salt, Pierre told Marie that she should select a name for it. She chose polonium after Poland, the country of her birth.[12] In December 1898, they isolated a second new element as a salt, and proposed the name radium for it.

Their discoveries were not immediately embraced by their colleagues, in spite of Pierre's professional reputation. No one at that time understood how an element, new or old, could spontaneously give off energy. The concept flew in the face of some of the most widely held theories of the time about the composition of matter. Neither physicists nor chemists were willing to accept a new theory until there was a convincing need for one. If there were indeed two new elements, what were their atomic weights? Could they be isolated in elemental form? Where did they fit in the existing periodic table of the elements?

If they were to isolate the new elements, the Curies would need more laboratory space. Across the courtyard from the room in which Marie had been working was a small shed. It had been formerly used by the medical faculty as a dissecting room, though it was no longer considered fit for even that function. For the next four years, the Curies worked in that small, drafty shed with its leaky skylight to extract and isolate the elusive elements. They even paid for their initial supply of several tons of pitchblende ore out of their own meager funds.

Pierre kept his full-time teaching position throughout the project. He helped with the physical labor when he could, but his teaching duties meant that Marie was forced to do most of it, as well as care for their eldest daughter, Irène. Marie shoveled tons of ore into large vats, then boiled, stirred, and poured it off into smaller containers for chemical treatment. It was backbreaking labor under the worst possible working conditions — stifling hot in the summer, freezing cold in the winter. Occasionally, they received help from a laboratory assistant from the school. In 1900, André Debierne, a young chemist at the institution, offered to help with their work. They asked him to search for yet a third new element that they suspected was present in certain rare clays. He succeeded in establishing its existence, and named it actinium.

It was not until 1902 that they were able to purify 100 mg of radium chloride and determine that radium's atomic mass was approximately 225. With this achievement their work became widely accepted by the scientific community. Like Roentgen, neither the Curies nor Becquerel would patent their work or take part in any commercial ventures. In 1903, the Curies and Becquerel shared the Nobel Prize for Physics in recognition of their discoveries.

Pierre was struck and killed by a horse-drawn cart on a Paris street in

1906. In 1911, Marie was awarded a second Nobel Prize, this time for chemistry, for the isolation of polonium and radium. She spent much of World War I traveling throughout France setting up X-ray facilities in hospitals for treatment of the wounded. She begged and borrowed cars and trucks which she outfitted with X-ray machines and dynamos to travel to hospitals where X-ray equipment was not available. These were known at the front as "Little Curies." During the course of the war, she established over two hundred X-ray facilities. In 1934, at the age of 66, Marie died of aplastic anemia, probably as a result of her exposure to radiation.

EARLY X-RAY RESEARCH

Roentgen was interested in X-rays from a purely scientific point of view. He wasn't interested in developing a practical use for them or in profiting from any such use that might be developed. This was not so for the rest of the world. Public speculation about practical uses began a scant five days after he announced his discovery. One of the reprints Roentgen mailed on New Year's Day went to his old friend Franz Exner, a professor of physics in Vienna. Exner, excited about his friend's discovery, promptly showed Roentgen's note and photographs to friends and colleagues at a party. One of the party guests showed them in turn to his father, who was the editor of the *Vienna Presse*. Its January 5th issue reported Roentgen's discovery and suggested that the X-rays might be useful to medicine.

> Biologists and physicians, especially surgeons, will be very much interested in the practical use of these rays, because they offer prospects of constituting a new and very valuable aid in diagnosis. At the present time, we wish only to call attention to the importance this discovery would have in the diagnosis of diseases and injuries of bones . . . The surgeon then could determine the extent of a complicated bone fracture without the manual examination which is so painful to the patient; he could find the position of a foreign body, such as a bullet or piece of shell much more easily than has been possible heretofore and without any painful examination with a probe . . . [8]

These words could hardly have been more prophetic. Though Roentgen himself expressed doubt that X-rays would ever be of practical use in medicine, he was besieged with letters asking for help with medical problems after his discovery was announced in major European newspapers. One letter, from a locksmith, told of his young son who had broken a leg. Though his injured leg seemed to heal, it continued to "get shorter" than his other leg. The father took the boy to a surgeon, who X-rayed the injured leg and told the father it would have to be rebroken to fit the ends together properly. The father asked Roentgen, "Wouldn't it be advisable if one leg had to be broken, to break the bones of the other leg in such a way that it would be shortened by the same length?" Although Roentgen consistently refused to answer any of these letters, he wrote on the margin of this one, "A rather workman-like idea of the locksmith. A physician would hardly have thought of that."[8]

U.S. Radiation Research

Word of Roentgen's discovery reached the U.S. in the first week of January 1896. By the end of that month, U.S. scientists had successfully made X-ray photographs. No one is certain who the first American to do this was, but convincing evidence exists that the honor belongs to Arthur W. Wright, a professor of astronomy at Yale University.[11] Wright, who was known on campus as "Buffalo" Wright because of his bushy beard, also had the distinction of being the first American to be awarded a PhD in science from an American university. On January 27, 1896, Wright made an X-ray exposure of a number of small objects, including a pencil and a pair of scissors. The following month, he made exposures of various animal bodies and a human hand. Whether or not Wright was in fact the first American to successfully expose an X-ray plate, he was a pioneer in the field, and the publication of his work led to scores of similar experiments.

Diagnostic Techniques Using Radiation

The first diagnostic use of X-rays in North America took place on February 3, 1896, at Dartmouth College. On that date, Edwin Brant Frost, a professor of astronomy, X-rayed the broken forearm of a boy who had been under the care of Dr. Gilman Frost (Professor Frost's brother) since the injury. The exposure clearly showed a healing fracture of the ulna.

X-rays were first used in conjunction with surgery in North America on February 7, 1896, when Professor John Cox of McGill University successfully located a bullet in a man's leg after several earlier attempts to locate it with a probe had been unsuccessful. Later that day the bullet was removed in surgery after being found exactly where Professor Cox predicted it would be.[11] This case is also notable because it was the first time that X-rays were introduced and accepted as courtroom evidence, when the patient successfully sued the man who shot him.

Despite these early medical uses of X-rays, they were not immediately accepted by the medical community. Many people, including scientists, were disturbed by seeing a skeletal image. Roentgen received many letters criticizing his experimentation with "death rays." Mrs. Roentgen was reported to have been unnerved by the sight of the bones in her hand and to have complained, as others did later, that it reminded her of death.[8]

A recent article reviewed the use of X-rays for diagnostic purposes between 1897 and 1927 at a Philadelphia hospital.[13] This hospital purchased its first X-ray apparatus in 1897, but a review of patient records for that year does not indicate that it was used at all. This may simply mean that the use of the machine was not recorded in patient records; however, this seems unlikely, since the X-ray machine was a new and undoubtedly exciting piece of equipment. Between 1897 and 1902, the machine was largely used out of curiosity rather than as a diagnostic aid. "Even a decade later, in 1912, roentgenographic examination was still quite an unusual event for the hospitalized patient. Patients with fractures severe enough to require surgery usually did not have a roentgenogram taken before going to the operating room for repair."[13] But by 1917, X-ray examinations had become routine for patients with kidney stones or fractures.

This hospital's use of X-ray technology may not be typical but nothing suggests that it is not. Physicians were undoubtedly as curious as anyone else about the new machine, but they had been diagnosing and treating bone fractures for thousands of years without any newfangled technology. Why leap to accept a new and virtually untested method of diagnosis until there was good reason to do so? Looking back at the problems of these early X-ray machines, particularly at the high doses of radiation emitted, the physicians of the time were undoubtedly wise in their conservative approach to this new technique.

X-Ray Therapy

Observations made in 1896 that X-ray exposures caused hair loss and skin burns led some therapists, though not necessarily physicians, to use X-rays in the treatment of hypertrichosis (excess hair growth), especially on the faces of women. The therapists noted that treated individuals who also suffered from acne saw an improvement in both conditions. Other investigators noted that X-rays appeared to be effective in the treatment of skin tuberculosis and chronic eczema. X-rays soon became a popular form of medical treatment. The December 1899 medical literature contains several reports of cases of skin cancer that were treated with X-rays.[11]

Radiation Therapy

Since X-rays seemed to be effective in treating certain tumors and skin diseases, radium was a natural choice to try in this role. At first, radium salts were taped directly to the affected skin area. But investigators soon discovered that radium's most common isotope (radium-226) emitted only alpha radiation, which was of little therapeutic value because it couldn't penetrate intact skin. Later, radium compounds were shaped into needles that could be inserted directly into target organs or body cavities through a hollow needle. One of the first conditions to be treated in this fashion was cancer of the cervix. The therapy proved to be useful, especially in inoperable cases.

Marie Curie discovered that the radium atom was transmuted (changed) into a new, gaseous entity (which she called "radium emanation") as it underwent radioactive decay. Radium emanation was officially recognized as an element in 1900, and was named radon in 1923.[14] In 1908, one of Marie's assistants developed an ingenious way to trap radon gas. The radon could then be sealed into glass ampoules, where its beta emission could radiate tissues.

Radiation Dosimetry

Measuring the radiation dose received by the patient was a difficult problem in the early days of radiation therapy. It was not until 1912 that a standard unit of radioactivity, the curie (Ci), was defined by international agreement as the number of radioactive disintegrations per second undergone by one gram of radium under specified conditions (37 billion). Other units of radiation dosimetry that took into account the types of tissue and radioactive emission (alpha, beta, or gamma) involved were adopted later.[11]

Radioisotope Tracers

Today, radioisotope tracers are commonly used in medicine. How these work is perhaps best explained in a story told by Nobel laureate George de Hevesy, who first used the technique. In 1923, Hevesy was living in a boardinghouse when he became suspicious that the proprietor was not serving fresh food as often as he claimed to. Hevesy brought a small speck of a naturally occurring radioisotope home with him and placed it on a piece of meat that he left on his plate. The next evening, Hevesy brought an electroscope (a radiation measuring device) to the dinner table and passed it over the main dish, which was hash. As he suspected, the hash proved to be radioactive. Hevesy did not mention what the response was from his fellow boarders. It's interesting to speculate on what their reaction would be if he pulled the same stunt today. Hevesy received the 1943 Nobel Prize for Chemistry for his study of plant physiology using radioisotope tracers.

Radioisotope tracers were first used in human clinical research in 1926, when three Boston physicians injected minute amounts of radium into a vein in a patient's right arm and measured the amount of time it took the radioactivity to reach the left arm. By performing the experiment on a large number of healthy people, they were able to establish a range of circulation times for healthy people. A circulation time longer than this range indicated impaired circulation.

ADVERSE HEALTH EFFECTS FROM RADIATION EXPOSURES

X-Ray Damage

Burns from X-ray exposures were reported very soon after Roentgen's discovery. The strength of X-ray emissions was dependent upon factors such as the thickness of the tube's glass walls and the amount of gas left in the tube by the inefficient vacuum pumps of the era. Many scientists constructed their own tubes, testing them by putting an object (usually their hand) between the energized tube and a film plate. A satisfactory exposure indicated that the tube was working properly. Because of the high doses of X-rays involved, skin burns were a common and accepted hazard of this procedure. In light of today's knowledge, it's easy to understand what caused the burns, but it was the subject of much debate at the turn of the century. Many theories were advanced in explanation. Some suggested that the burns were caused by ultraviolet light, while others believed that the burns were a result of the use of one source of current rather than another.[11] The prevailing attitude in 1896 is summed up in the words of one individual, who, after suffering severe burns and hair loss from X-ray exposures received while giving public demonstrations, said, "As to what produces the burn, I think it is purely an electrical effect, and that the ray has nothing to do with it."[15]

Machines that generated X-rays far more powerful than those Roentgen used became available soon after his discovery. Some researchers, in a desire to see more with the new rays, tinkered with different power sources, types of tubes, and longer exposure periods. These modifications, along with fact that the machines contained little or no shielding, exposed the operators to massive X-ray exposures. Exposures of hundreds or thousands of rems must certainly have been common. A 1902 literature survey found 55 cases of X-ray exposure injuries reported in 1896 alone.[11]

Some of the reported cases can be attributed only to extreme foolishness or stupidity. One glaring example was the case of a Connecticut physician who purchased an X-ray machine in 1899 and tested it by pointing it at his groin from a distance of five inches for 45 minutes. He suffered an intractable burn that required restorative surgery and a recovery period of a year and a half. Then he had the audacity to sue the machine's manufacturer for $20,000, alleging that he was injured because they falsely advertised that the machine would not cause burns. Even though the nation's foremost experts in the X-ray field testified on behalf of the company that the physician's actions were a gross misuse of the apparatus, he was awarded $6,750. Even by the standards of 1899, this "test" can be classified only as an act of self-mutilation or incredible stupidity. One wonders what a jury might have decided if the case were heard today.[11]

Possibly the first death as a result of radiation exposure was that of Friedrich Claussen of Germany. Claussen opened an X-ray laboratory in Berlin in February 1896, and gave hundreds of public demonstrations in which he X-rayed his own hands. He sustained severe burns on his arms that eventually required the amputation of his right arm. In 1900, he died of widely disseminated cancer.[16]

Clarence Dally

A well-known case is that of Clarence Dally, who was a glassblower in Thomas Edison's laboratory. Dally fabricated many cathode ray tubes for use in the lab, testing them in the usual way by exposing his hands. Edison was trying to replace the incandescent bulb with a fluorescent lamp made out of an X-ray tube with an inner lining of calcium tungstate. Dally fabricated a number of prototype lamps for this project. In 1896, Dally suffered severe X-ray burns of the hands from his repeated exposures. Although his physician warned him that if he did not give up his work his hands would have to be amputated, Dally continued to work. When one hand became too badly burned to use, he used the other. He was treated by a number of doctors between 1896 and 1904, and was diagnosed as having skin cancer around 1900. Recall that at that time X-rays were being used to treat skin cancer. Consequently, Dally was treated with a combination of skin grafts and *additional* X-rays. By the end of 1902, Dally had a large (3.5 × 2.5 inch) ulcer on his right wrist that had been treated with 144 skin grafts from his leg without success. In August 1902, his right arm was amputated at the shoulder. At that time, his surgeon urged him to have his left arm amputated as well, but Dally refused to give his consent. Later, he changed his mind, and his left arm was amputated above the elbow in March 1904. He died in October of that year.[17] Edison later claimed, and there appears to be no reason to question his statement, that when he began to suspect that the prototype fluorescent lamp was causing Dally's injuries he dropped the project.

Other radiation-linked deaths of the period have been tabulated from a number of countries. They occurred practically anywhere that X-ray research was carried out.[16]

Radiation Damage

It didn't take scientists long to discover that radiation caused skin burns. Becquerel suffered a burn soon after his discovery when he carried a piece of uranium salt in his waistcoat pocket for several days. Acting above and beyond the call of duty, Pierre Curie tried to duplicate Becquerel's results and did so successfully. The reports of these burns led others to perform experiments on animals, and also led to the idea that radium might be useful in treating certain skin conditions.

Radiation-induced genetic damage was first demonstrated by Charles Bardeen, a professor of anatomy at the University of Wisconsin, in 1907. Bardeen discovered that congenital defects could be produced in offspring by exposing the spermatozoa of toads to X-rays.[16]

The Dial Painters

Perhaps the single best-known case of radiation-induced cancer is that of the watch dial painters in the 1920s. The idea of using paints containing radium or thorium to make luminescent dials originated in Germany prior to World War I, and was used on German submarines during that war. Sabin A. von Sochocky, the originator of the idea, moved to the United States and became technical director of the U.S. Radium Corporation, which produced luminescent watch dials. The dials were painted primarily by young women capable of performing the necessary fine handwork. The women were not given any safety instructions for working with radioactive materials. The women worked out of small containers of paint which they held in their nondominant hand. They pointed their brushes by drawing them through their lips. Some historians have reported that some of the women found the radium paint so fascinating that they painted their teeth so that they would glow in the dark.[18]

In 1923, one of these women was diagnosed as having necrosis of the jawbone. The examining physician correctly diagnosed the cause and reported the case in the medical literature as "radium jaw." Two years later, the local medical examiner, Dr. Harrison Martland, became suspicious about the high mortality rate among the dial painters after nine of them died. Martland read the article on radium jaw, recognized the connection, and began what became a 20-year study of the dial painters, which he chronicled in a series of medical journal articles.

Incredibly, several years later, there were still scientists and physicians who argued vociferously against radium being the cause of the deaths, even though these events took place long after it was known that radium was deposited in bones and was capable of causing cellular damage.[19,20]

Ironically, in the course of his investigation, Martland became von Sochocky's personal physician, and signed his death certificate when he died of aplastic anemia in 1926.[16] An autopsy found the level of radium in his body to be greater than that of many of the dead dial painters.[21] After World War II, additional studies of the dial painters were funded by grants from the Atomic Energy Commission (AEC).

Many studies of the health effects of exposure to radiation have been conducted since the time of the dial painters. The use of atomic weapons and post-World War II atomic testing led to a number of studies of people exposed to radiation. A summary of all of these studies would be beyond the scope of this volume. Rosalyn Yalow reviews a number of them in Chapter 12.

RADIATION PROTECTION MEASURES

Measuring Radiation Doses

We'll never know for sure how many people died from the effects of acute or chronic radiation exposure in the early days of radiation research. We have no way of telling how high early radiation doses might have been, since there was no practical way to quantify the radiation given off by a source. By 1900, a number of scientists had developed methods to quantify radioactive emissions, but without a standard unit of measurement, the results of the different techniques were not comparable. One early method was to apply a barium platinocyanide coated patch to the patient's skin and estimate the radiation dose received from the color change of the crystals. However, this method wasn't very practical or accurate.

Development of the Film Badge

In 1907, Dr. Rome Wagner told an audience at an American Roentgen Ray Society meeting that he had begun carrying a photographic plate in his pocket while working with X-ray apparatus. Wagner's idea, now called a film badge, is still used to monitor the radiation exposures of radiologists, scientists, and other persons who work with radiation sources. Unfortunately, Wagner didn't think of the idea until after he developed cancer of the hands and cheek. He died of cancer only six months after describing his device.[16]

Early Radiation Protection Measures

The first statutory provisions for radiation protection were promulgated almost simultaneously in the early 1920s by the U.S., the United Kingdom, and Germany, at about the same time research began into the biomedical effects of ionizing radiation. The first standard developed was based on the amount of radiation necessary to produce the reddening of the skin (erythema) that Becquerel had so painfully discovered. This dose was called the skin erythema dose.

The International Unit of Radiation Dose

The first international agreement on a measure of radiation dose was signed in Stockholm, Sweden, in 1928. It defined and made the roentgen (R) the official unit of radiation dose. The International Council on Radiation Protection (ICRP) was established at the same meeting. The ICRP adopted (almost verbatim) the radiation protection recommendations of the British Protection Committee, since they were the only reasonably well-formulated policies in use at the time. These guidelines remained essentially unchanged until 1950.

In the U.S., a consolidation of several advisory groups took place in 1929, resulting in the formation of the Advisory Committee on X-Ray and Radium Protection. The group's name was changed to The National Council on Radiation Protection and Measurements (NCRP) in 1946.

The Radiation Tolerance Dose Philosophy

In 1934, both the American and International committees took the unprecedented step of establishing a "radiation tolerance dose" of 0.2 R/day (60 R/year), measured at the surface of the body. (Since the established U.S. practice was to measure radiation dose in air, the U.S. tolerance dose was adjusted to 0.1 R/day to maintain international equivalency.) The tolerance dose represented the level of radiation at which detrimental effects might be seen in the average individual. The committees recognized that any exposure to radiation might be detrimental, but felt that levels below the tolerance dose were generally safe and unlikely to cause harm in this average individual.[22]

In addition to their work in controlling the hazards of external radiation, radiation experts became concerned about internal body exposure to alpha radiation emitters, such as radium. In 1941, in the wake of the dial painters episode, the U.S. committee established a permissible body burden (level) of 0.1 microcuries (micrograms) of radium. These values became the baseline standards for radiation protection for Manhattan Project workers. In 1946, the NCRP increased in size and was divided into seven specialized committees. Its new members included representatives from the AEC, the military, and the Public Health Service.[23]

The Maximum Permissible Dose Philosophy

After its 1946 reorganization, the NCRP met frequently to review its prewar protection standards in light of the vast amount of radiation research done during the Manhattan Project. Partly in response to concerns about genetic risks, the NCRP decided to abandon the tolerance dose concept and adopt a "maximum permissible dose" standard. This action was taken to better convey the idea that there was no quantity of radiation exposure that was certifiably safe. The committee emphasized that to establish a permissible exposure dose, the dose limit must be based on the concept that " . . . the probability of the occurrence of such injuries must be so low that the risk would be readily acceptable to the average individual."[22]

In 1949, the committee decided to lower the permissible whole body dose limit to 0.3 R per six-day week (15 R/year). The action was taken because the committee wanted to maintain a conservative position, as they recognized that more people than ever before were exposed to more types of radiation as atomic energy research programs proliferated. Unfortunately, the NCRP position paper that explained this change was not published until 1954. This delay led to some controversy about the reasons for the changes, although the unpublished report had been widely circulated and debated in the scientific community.

The 1949 report was also significant because it contained the first recommended standard for nonoccupational radiation exposures. The nonoccupational dose was set at one-tenth of the occupational standard averaged over a full year. The new standard reflected both the NCRP's recognition of the sensitivity of children and the fetus to the effects of radiation, and that the public was typically exposed to radiation in small, infrequent doses.

The NCRP continues to review and recommend changes to existing radiation protection standards. Today, the permissible whole body occupational radiation exposure dose is 5 R/yr. Additional restrictions apply to certain isotopes, specific target organs, and a dose limit per calendar quarter.

ATOMIC ENERGY RESEARCH

The Discovery of Fission

The discovery of the neutron in 1932 stimulated radiation research in many new areas. In 1934, Irène Joliot-Curie (daughter of Pierre and Marie) and her husband, Frédéric Joliot, discovered that new radioisotopes were created when certain nonradioactive nuclei were bombarded with neutrons. This technique made possible the synthesis of nonnaturally occurring radioisotopes such as H-3 (tritium), for physiological research. The Joliot-Curies received the Nobel Prize for Chemistry in 1935 for this work. Within a year of their award, nearly 100 additional synthetic radioisotopes were produced by researchers around the world.[11]

In December 1938, two German physicists, Otto Hahn and Fritz Strassman, discovered that a small amount of barium was formed when uranium was bombarded by neutrons. Since the original uranium did not contain any barium, and barium's atomic mass (137) was about half that of uranium (238), the appearance of the barium was puzzling. Why was a lighter element formed when a heavy nucleus was bombarded with neutrons?

Hahn wrote of their discovery to Lise Meitner, a colleague who had fled Nazi Germany and was living in Sweden. Meitner, in collaboration with her nephew Otto Frisch, wrote a paper suggesting that the barium was formed by the splitting of the uranium atom into two lighter atoms. An American colleague suggested that the process be called fission, the term used for the splitting of living cells. Hahn received the 1944 Nobel Prize for Chemistry for the discovery of fission.

Frisch sent a copy of his paper to the famous Danish physicist Niels Bohr. When Bohr traveled to the U.S. in January 1939 to attend a scientific conference, he naturally carried this news with him. Fission was of great interest because it had the potential to release tremendous amounts of energy by the conversion of a tiny amount of matter to energy. A very small release of energy had been observed by Aston a decade earlier during the radioactive disintegration of uranium, but in theory, fission would liberate ten times more energy than radioactive decay.

The Self-Sustaining (Chain) Reaction

The concept of a self-sustaining or chain reaction was introduced by Hungarian physicist Leo Szilard in 1934. Szilard, who was Jewish, fled Germany for Great Britain when Adolf Hitler came to power in 1933. He suggested that a chain reaction might be initiated by bombarding heavy nuclei with fast-moving neutrons, and applied for a patent on a device that would use such a reaction, intending to turn the device over to the British government to forestall a similar move by the German government.

Szilard's idea would work only if the target nucleus emitted multiple neutrons when struck, a condition favored by the use of fast-moving neutrons. Neutrons emitted during radioactive decay were not energetic enough to reach this level. The search began for a source of fast moving neutrons.

Almost immediately, Enrico Fermi suggested that fission might generate neutrons energetic enough to support a chain reaction. He hypothesized that the higher ratio of neutrons to protons in heavy nuclei would make the nucleus unstable and increase the chance of a multiple neutron emission. He was absolutely correct.

Concerned that the Nazi regime would attempt to develop a fission weapon, Szilard organized a program of voluntary secrecy among physicists in the U.S. and Western Europe. In 1939, Szilard persuaded Albert Einstein to send a letter to President Roosevelt, warning him that atomic weapons were on the horizon.

The Manhattan Project

When Roosevelt received Einstein's letter, he convened a select group of scientists and top administration officials to secretly meet, determine whether other countries were developing atomic weapons, and consider the feasibility of the U.S. developing one. The committee's deliberations took on an air of urgency when Europe went to war in September 1939, and intensified after the Japanese attack on Pearl Harbor. In September 1942, the committee was dissolved when the Manhattan Project to develop an atomic weapon for the Allies[23] began. Major General Leslie

R. Groves, an army engineer, was placed in overall command of the project.

The project operated under conditions of unprecedented secrecy. Even the existence of the project was classified and, for security reasons, the project was referred to as The Manhattan Energy District. Because Groves bypassed the usual military chain of command and reported directly to the Secretary of War and the Chief of Staff, the project remained unknown to even the highest levels of the military, until the first atomic bomb was dropped on Japan.[24] This obsession with secrecy was not solely a result of paranoia. It was well known in scientific circles that several countries — including Germany, France, Japan, and the USSR — had been working on the development of atomic weapons prior to the war.

In retrospect, it is amazing that a project of this magnitude and importance could remain secret in wartime. Huge new research centers, virtually self-contained cities, were constructed in isolated areas of Tennessee, New Mexico, and Washington state. To give an idea of the scope of this empire, two plants were constructed on a 59,000-acre government reservation about 18 miles west of Knoxville, Tennessee. The new facility was named Oak Ridge, and once staffed, it immediately became the fifth largest city in the state.[25]

The project's enormous requirements for scarce construction materials such as cement and steel that were strictly rationed in wartime had to be obtained without revealing any details about the project. Though he carried a letter bearing the personal signature of the Secretary of War assigning the highest national priority to his project, Groves was constantly forced to fight to obtain required materials, equipment, and personnel.[24]

Construction of many of the project's facilities and much of its equipment was, of necessity, farmed out to firms in the private sector. Industrial giants, such as Dupont and Union Carbide, were asked to devote (and presumably risk) large amounts of financial and personnel resources to design and construct unique facilities for the project. Most of the firms quickly agreed to cooperate for the token compensation of a dollar a year.[26] Secrecy was maintained, in part, by having many firms work on small parts of the project without knowing the details of the overall effort.

The project made history on December 2, 1942, when the first controlled chain reaction was initiated in the world's first atomic pile (a primitive nuclear reactor), built on a squash court underneath the west stands of Stagg Field at the University of Chicago.

The first light of the atomic age came at 5:30 a.m. on July 16, 1945, in the desert outside of Alamogordo, New Mexico, when the world's first atomic bomb was detonated. The bomb was mounted on a tall, steel tower in a remote and heavily guarded corner of the Alamogordo air base. The test had been delayed for 90 minutes when local thunderstorms threatened to make air observation impossible. The closest observation point was the control site located 10,000 yards from the base of the tower. It consisted of a wooden bunker, reinforced with concrete and buried under massive layers of earth. At the moment of detonation, a blinding flash of light lit the sky enough to boldly outline a mountain range located three miles away. Seconds later, the light was followed by a tremendous roar and a pressure wave. (The shock wave knocked over several of the scientists who had left the shelter to watch the light effect, forgetting, in their excitement, to allow time for the pressure wave to pass.) Fortunately, no one was injured.[27] The now familiar mushroom cloud rose into the sky. When scientists inside a heavily shielded armored vehicle were able to reach the tower site, they found no trace of the tower—only a large crater where the desert sand had been fused into a green glass. The following month, the project culminated with the dropping of the first atomic weapons on the Japanese cities of Hiroshima and Nagasaki.

ATOMIC ENERGY REGULATION IN THE POSTWAR ERA

Creation of The Atomic Energy Commission

The Atomic Energy Act of 1946

Shortly after the end of the war, Congress held hearings on the peacetime regulation of atomic energy. Not surprisingly, the military felt that they should retain complete control over atomic energy. Their plans to retain postwar control of atomic energy within the War Department began as early as 1944.[22] The May-Johnson bill, authorizing the military to retain control over all atomic energy activities, was introduced in Congress almost as soon as the Japanese surrendered. The bill, however, ran into strong opposition from politicians and scientists who favored civilian control. In the Senate, a special committee was formed to hear testimony and recommend legislation.

After considerable debate during the spring and summer of 1946, a compromise bill, the Atomic Energy Act (AEA) of 1946, was passed and was signed into law by President Truman on August 1, 1946. This act created the AEC, which would function under the direction of five civilian commissioners to regulate atomic energy.[28] The AEC's principal functions were defined in the act as the production of fissionable material for weapons and the development and production of nuclear weapons to meet the needs of the military.[22]

The AEC faced the task of assuming responsibility for a project that had suffered a great deal in the immediate postwar period as its manpower and financial support were gradually withdrawn. The nearly one-year delay between the surrender of Japan and passage of the AEA forced the commission into an early decision with far-reaching consequences. They decided that the use of civilian contractors was the only feasible way to operate the project's facilities. Thus, though the government retained a monopoly on the control of atomic energy, private companies of necessity obtained knowledge and expertise by running government facilities.[22]

The Atomic Energy Act of 1954

The AEA of 1946 was unpopular with industrial interests that saw a large potential market closed off by its restrictive provisions. This group brought increasing pressure on both the AEC and Congress from 1946 to 1954 to permit the development of nonmilitary applications of atomic energy, such as commercial nuclear power plants and the use of radioisotopes in medicine and research. In response to these demands, Congress enacted the AEA of 1954. Chapter 1 of the Act declared:

> The development, use, and control of atomic energy shall be directed so as to make the maximum contribution to the general welfare, subject at all times to the paramount objective of making the maximum contribution to the common defense and security; and the development, use, and control of atomic energy shall be directed so as to promote world peace, improve the general welfare, increase the standard of living, and strengthen free competition in private enterprise.

The AEC's Successor Agencies

The AEC's responsibilities were assumed by two newly created agencies on January 19, 1975. The Nuclear Regulatory Commission (NRC) assumed regulatory responsibility for atomic energy activities, and the Energy Research and Development Administration (ERDA) took over atomic weapons research and development and research functions. ERDA became part of the newly formed Department of Energy (DOE) on October 1, 1977.

PUBLIC ATTITUDES ABOUT RADIATION

Pre-World War II Attitudes

The discovery of radioactivity was a tremendous thrill for scientist and citizen alike, as evidenced by the intense public interest about Roentgen's discovery. Scientists, promoters, and showmen by the score were willing to demonstrate the marvels of radioactivity to a fascinated public. X-ray demonstrations were very popular; they usually included X-rays of volunteers from the audience. While there is no question that substantial radiation doses were received by some volunteers, the net health effect was probably insignificant because they were one-time exposures that were unlikely to be repeated. Almost certainly the one at major risk was the person who put on the show.

Radioactive Nostrums

From the early years of the century through the 1920s, patent medicines, bottled waters, and prescription drugs containing radium and thorium isotopes were hawked by salesmen as cures for maladies such as rheumatism, hypertension, diabetes, and lagging sexual powers. A wide variety of pills, ointments, and salves containing radium were available in drugstores and by mail.[21,29] There was even a toothpaste containing radium to "brighten the teeth."[11] At least one company sold a device that added radium to tap water that was bubbled through it.[30] Many physicians routinely wrote prescriptions for radium or injected radium intravenously to treat diseases such as arthritis, gout, sciatica, and diabetes.[18] By the late 1920s, chocolate candy containing radium was on sale. Some physicians endorsed the use of these products; some routinely prescribed them for their patients. A citizen inquiry of the Bureau of Standards in March 1926, asked if the bureau had tested one of the products for the presence of radium. The next month, a citizen concerned about potential health effects contacted the Bureau and was told, "The Bureau has never heard of any cases of harmful effects due to drinking water which has been made radioactive."[21]

Most of these products were sold until they were banned by the federal government around 1932. The ban resulted from a federal investigation into the radium poisoning death of a prominent Pittsburgh steel manufacturer, who drank several bottles a day of water spiked with radium and thorium.[21] However, as late as 1953, one American company was still advertising a contraceptive jelly containing radium.[22]

By the late 1930s, public interest in radioactive materials had declined somewhat, due in part to government's increasing control over radioactive materials. Also, the radiation protection standards that had been promulgated on an international basis helped educate both scientists and citizens about the risks associated with radiation exposures. From the start of the war in Europe until the atomic bombs were dropped on Japan, the public was to hear little about radioactivity.

Atomic Weapons

Hiroshima and Nagasaki

The first atomic weapon used was a 20-kiloton atomic bomb dropped over Hiroshima, Japan, on August 6, 1945. The White House released the news at midmorning of what was otherwise a slow news day. Early damage estimates were scanty because the immense cloud of dust from the explosion obscured the view of the observation plane. It was clear, however, that the city had been largely destroyed. President Truman was returning from the Potsdam Conference aboard the U.S.S. *Augusta* when he received word of the successful bomb drop. He rushed to the officers' wardroom to spread the news, where it was greeted with loud cheers.[30]

Much of the nation heard the news on noon radio broadcasts or by word of mouth. It came as a complete shock to most because of the secrecy surrounding the project. Even Mrs. Leslie R. Groves told reporters, "I didn't know anything about it until this morning, the same as everyone else." After the initial announcement, journalists all over the country worked to find a new angle for the story. In Hanford, Washington, local residents expressed surprise that the new top secret base had been producing plutonium; they had assumed all along that it was making poison gas.[30]

On August 9th, just three days after the Hiroshima bombing, a second atomic weapon was dropped, this time on the Japanese city of Nagasaki. Again, the city was virtually destroyed. The next day, the Japanese government offered to surrender if Emperor Hirohito could maintain his throne. The Allies agreed, and on August 14th, the war was over, though the surrender papers were not signed until the 17th, in a ceremony on board the U.S.S. *Missouri* in Tokyo Bay.

After the War

The public's reaction to the use of atomic weapons was first shock, then joy. It was clear to all that the Japanese surrender was a result of the destruction wrought by the bombs and the U.S. threat that more would be dropped (an empty threat, since no other bombs existed). Allied military strategists had been planning an invasion of Japan right up until the day of the Hiroshima bombing. Expecting fierce opposition, the predicted casualty tolls for the invasion were in the hundreds of thousands. Not unreasonably, a nation tired of war (Germany had surrendered three months earlier) embraced any weapon that would bring an early end to the fighting.

Life magazine devoted much of its August 20, 1945, issue to the bomb. The issue included full-page photographs of the two destroyed cities and a first look at the ominous mushroom cloud that followed the explosions. In the same issue, an editorial stressed the point that atomic fission was a major scientific breakthrough that would enable mankind to understand and tame nature.

After the initial shock and jubilation wore off, Americans began to show the same sort of enthusiasm they demonstrated when Roentgen discovered X-rays. Businesses were quick to jump on the bandwagon. Within hours of the White House announcement, the Washington Press Club offered an "atomic cocktail" of Pernod and gin. Department stores advertised "atomic sales," and even the staid *New Yorker* ran some of a multitude of jokes revolving around the word atomic. Some entrepeneurs, ignoring the threat of radioactivity, gathered fused green-colored sand from the Alamogordo test site and fashioned it into jewelry, which they sold nationally. The General Mills Corporation offered an Atomic Bomb Ring, complete with bombardier's insignia, a secret message compartment, and a sealed atom chamber where one could see atoms split to "smithereens." Their advertising copy said that the ring worked on scientific principles used in the laboratory to study nuclear fission, and guaranteed that any child could wear one in complete safety. All this excitement was available for only 15 cents and a boxtop from Kix cereal. Some 750,000 youngsters deluged the company with orders.[30]

Other products capitalized on the public's fascination with the bomb. Books and movie scripts were altered to get some mention of atomic weapons into their titles or plots. A new magazine, *The Atom*, was announced by the Atomic Age Publishing Company of Denver. The 1947 Manhattan telephone directory listed 45 businesses with the words atom or atomic in their names (including The Atomic Undergarment Company).[30] Songs with titles such as "Atom Cocktail," "Atom and Evil," and "When the Atom Bomb Fell" were recorded between 1945 and 1947. One California record company marketed jazz records on the "Atomic" label, adorned with a picture of a mushroom cloud.

Not everyone was so enthusiastic, however. Hanson W. Baldwin wrote in the *New York Times* on the day after the bombing of Hiroshima, "The bomb might indeed force Japan to her knees but it would also bring incalculable new dangers . . . We have sowed the whirlwind."[31]

Some Manhattan Project scientists had deeply ambivalent feelings about working on a weapon of mass destruction and were adamantly opposed to actually using the atomic bomb. A number of them lobbied the Administration through the spring and summer of 1945 urging that the bomb not be used until a demonstration of its power could be arranged for the Japanese leadership and the world press. They were convinced that the Japanese would surrender if the power of the bomb was demonstrated.

Postwar Attitudes

In the immediate postwar euphoria, the media was full of fanciful descriptions of a future with atomic power. Science writers speculated freely about the potential uses of atomic energy. Atomic powered cars and airplanes would travel "for a year on a tiny pellet of uranium." Artificial suns would enable man to control the weather; the polar ice caps could be melted to provide water for desert regions of the world.

By late 1945, cooler heads were cautioning against these daydreams. In November 1945, Albert Einstein stated that "no practical benefits (of atomic energy) would be available for some time." Arthur Compton pointed out that the massive lead shielding required by nuclear reactors would allow only stationary power plants to be plausible. In August 1948, *Time* magazine asked, "What happened to the nation's glamorous program for peacetime uses of atomic energy?"[30]

Part of the answer, of course, was that the military was still holding very tight control over atomic energy. The only research being done in this field was sponsored by the military, and they weren't interested in atomic cars.

Changing Times

What changed the public's attitude about atomic energy from wildly speculative optimism to deep-seated fear? Undoubtedly, a number of factors were at work. First, the same prominent scientists who had opposed the use of the bomb continued to be publicly critical of its further development. They worked to counteract what they saw (undoubtedly correctly) as a government propaganda campaign to promote the continued development of atomic weapons. Second, when the immediate postwar euphoria came to an end, many people had second thoughts about what future wars would be like with atomic weapons. Third, the Cold War had increased international tension, and the idea of a third World War no longer seemed farfetched.

The AEC and the commercial firms involved in atomic research worked hard to comfort those who worried about atomic energy. The AEC established a Division of Public and Technical Information in October 1947, to function essentially as a public relations department. In a pilot project, which grew into a national campaign in 1948, the AEC and its contractors sponsored a series of Atomic Energy Weeks around the country. These were complete with lectures by prominent scientists, elaborate exhibits, and characters such as Miss Molecule and Mr. Atomic Energy. The AEC estimated that four million people attended such exhibitions by the end of 1948. Brookhaven National Laboratory (BNL) was a particularly prominent participant in these activities because it was the site of much of the AEC's nonmilitary research.[30]

As the Cold War heated up and the USSR detonated its first atomic bomb (constructed as a result of intelligence gathered from western nations), the government's emphasis shifted from Miss Molecule to teaching the public how to build fallout shelters. Civil defense shelters, stocked with food, water, medicines, and portable toilets, were built in public buildings. Millions of pamphlets explaining how to turn a home basement into a fallout shelter were sent home with children or mailed out by government agencies upon request. Many of us remember the Cuban missile crisis of 1962, when we feared those shelters might be needed. Those who were in school during the crisis will undoubtedly never forget the "shelter drills" where we sat in a row, hugging our knees to our chests, in a basement or ground floor hallway, with our shopping bag of canned goods beside us. Though the government cancelled its fallout shelter program years ago, and 25 years have passed since the Cuban missile crisis, some of the public's negative attitudes toward atomic energy clearly date from these days.

RADWASTE DISPOSAL

Early Practices

Atomic energy research increased sharply after World War I. Public interest was high, research funds were easy to obtain, and restrictive regulations on handling radioactive materials had not yet been promulgated. Radioactive waste (radwaste) disposal was not viewed as a serious problem because waste volumes were small and the scientists handling them were perceived to have the knowledge and wisdom required to handle them safely. In most cases the wastes were burned, buried in shallow trenches, allowed to evaporate, or poured down the sewer at the point of generation. The "dilute and disperse" philosophy (i.e., minimize the radiation dose to individuals by diluting and dispersing the radioactive material) underlies each of these methods.

Manhattan Project Wastes

The Manhattan Project generated far greater volumes of radwastes and at faster rates than ever before. New, transuranic radioisotopes were generated as byproducts of the operation of atomic reactors. Methods were developed to dispose of the various types of radwastes generated by the project. Some of these methods included open burning, shallow land burial (SLB) onsite, closed incineration, evaporation, pouring wastes into the sewer system, or storing them for later disposal. Major disposal sites for the project were opened at the facilities in Hanford, Idaho Falls, Los Alamos, Oak Ridge, and Savannah River.

How the government disposed of radwastes from the project should be viewed in perspective. First, those radwastes were not classified in the same way as they are today; the definitions of high- and low-level wastes have changed over time. Currently, low-level wastes are defined by statute as wastes that are not high-level wastes. However, at the time of the Manhattan Project, many types of wastes that were handled as low-level wastes would today be considered high-level. Second, it is understandable that during the war, waste disposal was never a high priority; obviously, all available resources were needed to work on the project's main thrust. Unfortunately, even after the war, the AEC failed to devote significant resources to either waste disposal research or the development of a national radwaste policy. No one is sure how much radwaste from the Manhattan Project is still stored around the country. However, it is safe to say that today, 40 years after the end of the war, thousands of gallons of liquid radwastes and many tons of solid radwastes from the project are still stored around the country.

The reasons for the commission's failure to develop a national policy on radwaste disposal are easier to understand than they are to excuse. Let's face it: working with garbage, radioactive or otherwise, has never led to public acclaim or recognition in the scientific field. The radwaste disposal problem is almost exactly analogous to the chemical waste disposal situation, where there were no federal, and few state, laws regulating chemical waste disposal until 1976. In that year, Congress passed the first comprehensive law regulating chemical waste disposal, in the aftermath of publicity about Love Canal. This legislation was passed in an atmosphere of fear and anger over the reports of illnesses and injuries suffered by residents in the area. Neither state nor federal agencies were able to deal effectively with the public's fears about the situation. Why did this situation arise? It is important to remember that the disposal of wastes at Love Canal was done in complete accordance with the laws and practices of its day.[32] Everyone dumped chemical wastes on the ground or in pits as a simple matter of economics; no firm could afford to spend any more money than necessary handling either chemical wastes or industrial refuse. Since no one saw chemical waste disposal as a problem, no one bothered to regulate it.

Postwar Disposal of Radwastes by the AEC

When the AEA of 1946 gave the AEC statutory authority over all aspects of atomic energy, it became legally responsible for both high- and low-level radwaste disposal. The agency's primary method of disposal was SLB at AEC disposal sites on federal lands. SLB techniques developed in a haphazard fashion from the shallow trenches used by early researchers. Generally, trenches (of various depths) were dug onsite, the radwastes were tossed into them, and the trenches were filled in with the excavated dirt. Because some wastes were not containerized, spectacular chemical reactions or fires occasionally occurred when incompatible compounds were mixed. SLB continues to be the most common method of low-level radioactive waste (LLRW) disposal today.[33]

Ocean Disposal

Large-scale ocean disposal was first used by the U.S. in 1946, when the AEC arranged to have the Navy dump radwaste containers at sea. About 95% of the containers were 55-gallon steel drums, filled with concrete after wastes were placed in them.[34] Bulky pieces of apparatus and equipment were placed in concrete-filled wooden boxes or concrete tanks.[35] The containers were then hauled out to sea and dumped from ship or barge decks. Most of these wastes were disposed of at sites off the Atlantic and Pacific coasts; small amounts were dumped into the Gulf of Mexico.[34] Figures 1–8 show what these operations looked like. Some Manhattan Project wastes were disposed of in this manner. Civilian contractors were licensed to perform the dumping for the AEC after the Navy was used for several years.

Between 1946 and 1967, the U.S. alone dumped some 90,000 radwaste containers into the oceans. Ocean disposal was slowly phased out (by the

Figure 1. Dry radioactive wastes being packaged in a 2.5 gallon cardboard container prior to placement in drums. Note film badge on lapel of lab coat and survey instrument on table. (Source: U.S. AEC.[40])

Figure 2. Primary containers being loaded into steel drum which is partially filled with concrete. Note unbuttoned lab coat and bare arms between sleeve and gloves. (Source: U.S. AEC.[40])

Figure 3. Drums waiting to be filled with concrete. Full drums in background. (Source: U.S. AEC.[40])

Figure 4. Staging area for drums prior to loading on barge. (Source: U.S. AEC.[40])

Figure 5. Arranging drums on deck for ready jettison. Film badge is visible on lapel of individual in dark jacket. (Source: U.S. AEC.[40])

Figure 6. Tug is pulling barge in a circular pattern over dumping area. Securing lines are being released to allow drums to roll off deck. (Source: U.S. AEC.[40])

Figure 7. Drum rolling off deck. Readying drums on far right for release. (Source: U.S. AEC.[40])

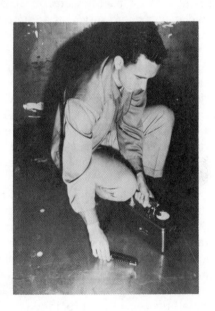

Figure 8. Checking barge deck for residual contamination with survey instrument. (Source: U.S. AEC.[40])

U.S.) in the early 1960s when it became more expensive than land disposal.[36] Only 36 radwaste containers, containing a total activity of three curies, were dumped by the U.S. in the ocean in 1970. The practice was formally ended by the U.S. in June 1970. In 1972, Congress placed ocean dumping of radwastes under the jurisdiction of the EPA.[30,37] No country is currently dumping radwaste containers at sea, but some European countries still dispose of radwastes from nuclear power plants by piping them offshore into their coastal waters.

The environmental impact of ocean dumping of radwastes has been hotly debated. Ocean dumping, like other socioscientific controversies, often boils down to the diametrical conclusions reached by critics and proponents of the technique when they view the same research results. To cite one example, the basic assumption behind ocean dumping was that the radioactive material would be diluted and dispersed by the ocean's great volume of water before anyone was exposed to it. This is the same concept under which the NRC allows sewage disposal and incineration of specified types and quantities of radwastes.

In the case of ocean dumping, several studies have been conducted by government agencies and individual researchers. Those who oppose the practice claim that radionuclides have been released by damaged containers and have bioaccumulated in the food chain. They cite a 1976 EPA underwater photo survey of a dump site near the Farallon Islands (off the California coast), which showed that one quarter of the drums photographed had ruptured. A concurrent study of sediment in the area reportedly found plutonium levels to be elevated significantly above levels that might be expected from fallout. In a follow-up study, one university scientist claimed he found elevated levels of radioactivity in fish caught off the California coast from 1974 to 1976.[38] On the other hand, a 1984 Government Accounting Office (GAO) report states that the environmental hazards of ocean disposal of radwastes have been overemphasized, and found no evidence of bioaccumulation of radionuclides in the food chain.[36]

Unquestionably, many of the dumped containers were damaged in their fall to the ocean floor or have been breached since they were dumped. However, many of them appear to have survived in fair shape. In 1976 an 80-gallon steel drum was recovered from a depth of 9,000 feet in the Atlantic Ocean. It had been on the ocean floor for about 15 years prior to its recovery. The drum was taken to BNL for detailed examination. Though the drum had major indentations from the pressure differential, the concrete matrix inside the drum was intact.[39]

Disposal Practices at Oak Ridge National Laboratory (ORNL)

Land Disposal

Generally, the Hanford, Los Alamos, and Savannah River sites were used only for AEC wastes.[40] ORNL and the National Reactor Testing Station (now the Idaho National Engineering Laboratory) were the two principal sites where commercial LLRW was handled until the AEC stopped accepting it in 1962. The disposal techniques used at ORNL have been well documented and probably represent a typical example of disposal practices in the 1940s and 1950s.

SLB, used from the beginning of the ORNL operation, was unquestionably the most commonly used technique. Different procedures were used for different radionuclides. Alpha emitters were placed in pits and covered first with a foot of earth, then with 18 inches of concrete to prevent accidental future intrusions; after this, additional dirt was added to bring the cover to ground level.[35] Beta- and gamma-emitting wastes were placed in trenches 200 feet long, 10 feet wide, and 15 feet deep. Wastes were dumped in one end of the trench and covered by 3 or more feet of earth. Extremely radioactive materials were dropped into 15-foot-deep augured holes to minimize the exposure of site personnel.[35] The early disposal sites were relatively small and poorly located without benefit of geologic exploration.[40] Though records were kept of the contents of each trench, many of these records were reportedly destroyed in a fire in 1959.[38]

Figures 9–12 illustrate an unusual case of the disposal of solid radwastes. They were taken at the Los Alamos National Laboratory in early 1954, during the demolition and disposal of a 48,000 square foot building that was badly contaminated with plutonium. (The building had been used for plutonium production from the beginning of the Los Alamos project.) Plutonium is an alpha-emitting radioisotope. Alpha particles cannot penetrate intact skin, but are damaging to internal tissues if inhaled, swallowed, or allowed to penetrate the skin. SLB was selected as the disposal option because the building contained large amounts of noncombustible material. A pit capable of holding 216,000 cubic feet of waste material was prepared at a disposal site seven miles from the demolition site.

The laboratory's Health Division decided that workmen should wear air-purifying respirators during the demolition operations because of the inhalation hazard of the dust. To contain the alpha contamination, the building was sawed into 8 × 10 foot sections and lifted by a crane onto a truck. Before sawing, each section was thoroughly soaked with water from a fire hose. Once loaded on the truck, the sections were covered with cheesecloth, sprayed with paint (to minimize dust), and then wrapped with canvas. Over 1,000 truckloads were hauled to the disposal site.[40]

Liquid wastes were more difficult to handle than solids because they were more mobile and represented a threat to ground water. The original ORNL liquid waste disposal system consisted of eight underground tanks with a combined capacity of one million gallons. The tanks were unlined;

Figure 9. Building before start of demolition operations. (Source: U.S. AEC.[40])

Figure 10. Crane removing section of building. Note individual with water hose spraying section to reduce dust. (Source: U.S. AEC.[40])

any discharge from them could (and soon did) drain almost directly into the stream which provided the water supply for the local community.

Some liquid wastes were solidified with cement or absorbed on solids before being placed into disposal trenches. This technique, however, required greater volumes of disposal space than did storage in tanks, so plans were made to add an extra 300,000 gallons of storage tank capacity. While these tanks were being built, liquid wastes were disposed of in open pits as a "temporary" measure.[40] Liquid radwaste disposal pit #1 was opened until the new tanks were ready. The pit was 100 feet long, 20 feet wide, and 15 feet deep, giving it a capacity of 180,000 gallons. It was roofed to prevent it from filling with rainwater, and wire mesh screening was later placed over its surface to prevent wildfowl from landing on it.

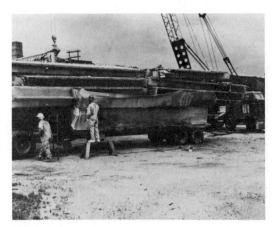

Figure 11. Section has been loaded on truck and is being wrapped in cheesecloth and sprayed with paint. Note all individuals are wearing respirators. (Source: U.S. AEC.[40])

Figure 12. Loaded truck with cargo wrapped with canvas. Although obviously a posed shot, note that one man is sitting on top of the radioactive waste material and the driver's respirator is hanging around his neck instead of covering his mouth and nose. (Source: U.S. AEC.[40])

Figures 13 and 14 show the pit being excavated and its appearance after the roof was added.

These pits contained a radioactive soup of aqueous and organic chemical solvent wastes. Radiation levels around the perimeters were measured in 1956 and found to represent a hazard to workers. The permissible occupational dose rate at that time was 7.5 millirems/hour. One pit had a radiation level at its edge of 1000 millirems/hour. At that dose rate, an individual could remain at the edge of the pit for only 27 seconds out of each hour without exceeding the occupational exposure limit.[35] One gets the image of men running up to the edge of the pit to dump waste and then sprinting back to safety.

Two and a half months after the pit was opened, after 123,000 gallons

Figure 13. Liquid radioactive waste pit being excavated in the early 1950s at Oak Ridge National Laboratory. (Source: U.S. AEC.[40])

Figure 14. Finished pit fitted with roof and wire mesh to keep rain and animals out. (Source: U.S. AEC.[40])

of waste had been placed in it, radioactive contamination was detected 150 feet downhill from it. The pit was then taken out of service. Other pits were also constructed, but either they leaked or seepage of ground water into them reduced their storage capacity.

By 1951, ORNL was being overwhelmed with both high- and low-level radwastes from a number of sources. Radwaste burial rates tripled between 1945 and 1955. Most of the increase was a result of the large amount of research being done for the AEC. (Until the AEA of 1954, atomic energy research was conducted only under government contract.) By 1955, 25 different institutions or government agencies were using Oak Ridge as a disposal site.[40]

State Radiation Protection Programs

The AEA of 1954 was amended in 1959 to allow states to administer their own radiation protection programs. State programs were required to be at least as stringent as the federal regulations, and were required to be reviewed and approved by the AEC before they became effective. States that received AEC approval are known as "agreement states."

In July 1961, Kentucky became the first state to apply for agreement state status. In its application, Kentucky requested permission to regulate the disposal of LLRW within its borders. When the AEC processed the application, its staff opposed granting the state authority to regulate LLRW on two grounds. First, they were uncertain that individual states were qualified to provide long-term supervision of waste sites. Second, the agency felt that continued federal control was advisable because waste disposal could be most efficiently managed on a regional basis.[41] One farsighted AEC official predicted that, "If states assume jurisdiction in this matter (LLRW) each state would want a burial site within their borders."[22] As we shall see in the next chapter, this was hardly the case. Kentucky was granted agreement state status in February 1962, without being granted authority over LLRW disposal.

Kentucky was not alone in its quest for LLRW control; in 1962, the Southern States Governors' Conference also voiced objections to the AEC's absolute control over LLRW disposal practices. Furthermore, private industry, which saw a potentially lucrative market developing, had been pressuring the agency to allow the operation of commercial waste disposal sites since at least 1960.[23] In the face of sharply increasing waste volumes, pressure from the states and private industry, and a tight budget, the AEC began to reconsider its position.

Commercial Radwaste Disposal

In 1962, the AEC announced that it would accept license applications from private companies that wished to operate LLRW disposal sites. The sites were to be owned by the states in which they were located and operated by commercial firms under federal (or agreement state) regulatory programs. The first such site opened in Beatty, Nevada, later that year; the second opened the following year in Maxey Flats, Kentucky.

Now, with two commercial radwaste disposal sites open, the AEC took a step which would have far-reaching consequences. They decided that since the two commercial sites were adequate to handle all of the commercial LLRW then being generated, they would stop accepting commercial LLRW for disposal. The commission did, however, continue to operate disposal sites for wastes from weapons production, and to dispose of commercial high-level radioactive wastes (HLRW) at its Hanford, Washington, site. As we shall see in the coming chapters, this seemingly innocuous decision is what led to this book.

In 1971, the commercial radwaste disposal business reached its peak when the sixth (and last, to date) site was opened in Barnwell, South Carolina. By 1979, only three of the six were still open (and one of those, in Beatty, was scheduled to close within six years). One of the sites (Sheffield, Illinois) closed when it was filled; the other two (Maxey Flats, Kentucky and West Valley, New York) were closed because of technical difficulties in their management. As a result, in 1979, the governors of the three states with operating sites threatened to close them if they were forced to continue taking all of the nation's radioactive trash (two of the sites were closed for short periods that year, partly to reinforce the governors' points).

This situation provided the impetus for congressional involvement in LLRW in the 1980s. The story of this activity is told in Chapter 2.

REFERENCES

1. Geissler also invented the mercury pump that made it possible to evacuate the tubes. The efficiency of the vacuum achieved was not impressive by today's standards, but it sufficed for the day.[3]

2. Asimov, I. *Worlds Within Worlds: The Story of Nuclear Energy, Vol. 1* (Washington, DC: United States Atomic Energy Commission, 1972).

3. Donzetti, P., translated by A. Ellis. *Shadow and Substance; The Story of Medical Radiography* (New York: Pergammon Press, 1967).

4. Crookes also discovered and isolated the element thallium and determined its atomic weight. He also invented the radiometer, which converts radiant energy into rotary motion, and demonstrated that the emission spectrum of helium on earth was identical to spectra observed from the sun, thus proving that helium was present in the sun.[3]

5. Many literature references from the period refer generically to "Crookes" tubes, leaving poor Geissler something of a nonentity. The mistake is still common today.

6. The correct spelling of his name is Wilhelm Conrad (not Konrad, a common mistake) Röntgen. I have chosen to anglicize it because the historical unit of radiation dose was the roentgen.

7. Crookes actually produced X-rays before Roentgen did but did not realize it. He found some fogged photographic plates in his laboratory and wrote a letter of complaint to the manufacturer. He did not realize the significance of the incident until Roentgen announced his discovery.

    An X-ray exposure of a photographic plate was made by physicist Arthur

W. Goodspeed and photographer William N. Jennings on February 22, 1890. The exposed plate was placed into a drawer of other unexplained photographic phenomena and forgotten until after the announcement of Roentgen's discovery. This plate exists today. To their credit, none of the above ever claimed to have beaten Roentgen.

The word *radioactivity* was not used until Marie Curie first proposed it in 1897, during her search for elements other than uranium that released energy spontaneously. Before that, the emissions were known as X-rays (or Roentgen's rays) or Becquerel's rays.

8. Glasser, O. *Dr. W. C. Roentgen*, 2nd edition (Springfield, IL: Charles C. Thomas Inc., 1958).

9. Among other discoveries, Lenard demonstrated that the power to absorb cathode rays was dependent on density, rather than chemical nature. He also worked extensively on the study of phosphorescence and was awarded the Nobel Prize for Physics in 1905.

10. This particular tube *was* one of Crookes's designs.

11. Brecher, E. and R. Brecher. *The Rays: A History of Radiology in the United States and Canada* (Baltimore: Williams & Wilkins Inc., 1969).

12. Curie, E. *Madame Curie* (New York: Doubleday & Co. Inc., 1937).

13. Howell, J.D. "Early Use of X-Ray Machines and Electrocardiographs at The Pennsylvania Hospital," *J. Am. Med. Assoc.* 255(17):2320–2323 (1985).

14. Radon has, of course, been in the news a great deal in the last few years, after being detected in homes at levels that some scientists consider unsafe. Its presence in houses is explained by the fact that radium is a common element in certain geologic formations, such as granite deposits. As the radium in bedrock underlying a structure undergoes radioactive decay, it is transmuted to radon gas, which enters the structure mixed with air. Although this phenomenon has been occurring since the beginning of time, indoor radon levels have increased as "tighter" buildings are constructed to save energy.

15. Codman, E.J. "The Cause of Burns From X-Rays," *Boston Medical-Surgical J.* 135:610–611 (1896).

16. Grigg, E.R.N. *The Trail of the Invisible Light* (Springfield, IL: Charles C. Thomas Inc., 1965).

17. Brown, P. *American Martyrs to Science Through the Roentgen Rays* (London: Bailliere, Tindall & Cox Inc., 1936).

18. Pringle, P. and J. Spigelman. *The Nuclear Barons* (New York: Holt, Rinehart and Winston, 1981), p.187.

19. Castle, W.B., K.R. Drinker, and C.K. Drinker. "Necrosis of the Jaw in Workers Employed in Applying a Luminous Paint Containing Radium," *J. Ind. Hyg.* 7(8):371–382 (1925).

20. Flinn, F.B. "Radioactive Material An Industrial Hazard?" *J. Am. Med. Assoc.* 87(25):2078–2081 (1926).

21. Bartlett, D.L. and J.B. Steele. *Forevermore: Nuclear Waste in America* (New York: W.W. Norton & Company, 1985).

22. Mazuzan, G.T. and J.S. Walker. *Controlling the Atom: The Beginnings of Nuclear Regulation 1946–1962* (Berkeley, CA: University of California Press, 1984).

23. Actually, the intent was to develop an atomic weapon for U.S. use on behalf of the Allies. When the project began, high officials of the British government agreed to drop their own nuclear research program (code named Tube Alloys) to avoid duplication of effort and to maximize efficiency. The agreement was that after the war, the U.S. would share research results with the cooperating governments. After the war, the McMahon-Johnson bill, placing atomic energy under civilian control and prohibiting the disclosure of information to any foreign power, became law. The English were rightfully bitter over this and it took more than a decade for the damage to Anglo-American relations to be repaired. The English proceeded to develop atomic energy on their own. [Hyde, H.M. *The Atom Bomb Spies* (New York: Ballantine Books, 1980), pp. 48–49.]

24. Groves, L.R. *Now It Can Be Told: The Story of the Manhattan Project* (New York: Da Capo Press, Inc., 1962).

25. "Atomic Energy Harnessed In Great Scientific Achievement," *Chem. Eng. News* 23(15):1401–1408 (1945).

26. In a routine postwar audit of government accounts, the Dupont company was actually asked to rebate less than a dollar, because the war ended in August 1945, and the company had already been paid its dollar for the year. Fortunately, the Dupont people showed a sense of humor; they repaid the money and had a well-deserved laugh at the inanities of bureaucracy.[24]

27. Lawrence, W.L. *Dawn Over Zero: The Story of the Atomic Bomb*, (New York: A. Knopf, Inc., 1946).

28. Other provisions of the Act retained exclusive government authority over all development and applications of atomic energy, a government monopoly on the ownership of fissionable materials and the facilities that produced them. It tightly controlled any existing or future patents obtained from government research and placed sharp restrictions on cooperation with other countries in atomic energy projects. The latter provision angered the British, who felt it violated the wartime agreement that England would drop its own efforts to develop an atomic weapon and cooperate with the Manhattan Project in exchange for information access after the war.

29. Eisenbud, M. *Environmental Radioactivity* (New York: Academic Press, 1973).

30. Boyer, P.S. *By The Bomb's Early Light* (New York: Pantheon Books, 1985).

31. Baldwin, H.W. "The Atomic Weapon," *New York Times*, August 7, 1945, p. 10.

32. Burns, M.E. "Striking A Reasonable Balance," in *Hazardous Waste Management: In Whose Backyard?*, M. Harthill, Ed. (Boulder, CO: Westview Press, Inc., 1984), pp. 185–200.

33. "The Problem of Disposing of Nuclear Low-Level Waste: Where Do We Go From Here?," U.S. General Accounting Office, (1980).

34. "Radioactivity in the Marine Environment," National Research Council (1971).

35. Straub, C.P. *Low-Level Radioactive Wastes: Their Handling, Treatment and Disposal* (Washington, DC: U.S. Atomic Energy Commission, 1964).

36. "Hazards of Past Low-Level Radioactive Waste Ocean Dumping Have Been Overemphasized," U.S. General Accounting Office Report EMD-82-9 (October 1984).

37. "Managing Low-Level Radioactive Wastes: A Proposed Approach," The National Low-Level Waste Management Program, U.S. Nuclear Regulatory Commission (1980), pp. 10–11.

38. Shapiro, F.C. *Radwaste* (New York: Random House, 1981).

39. Short, H. "Sea Burial of Radwaste: Still Drowned In Debate," *Chem. Eng.* 91(5):14–18(1984).

40. "Sanitary Engineering Aspects of the Atomic Energy Industry," U.S. Atomic Energy Commission Report TID-7517 (October 1956).

41. This was the first mention we've found of the concept of managing radwastes on a regional basis rather than turning them over to the government for disposal. Although the AEC's disposal sites were located in almost every region of the country, only the sites at ORNL and NRTS (Idaho) were used for the disposal of commercial radwastes (with a few exceptions). The regional disposal site concept sounds attractive but there is a basic flaw in it. Not all areas of the country have the right combination of geology and weather to make them suitable for SLB. An ideal site would be in an area of low population density with a deep water table and low annual rainfall. A poor area would have exactly opposite characteristics. Although it might not be "fair," the western part of the country is much closer to the ideal than the eastern part.

CHAPTER 2

The Politics of Low-Level Radioactive Waste Disposal

This chapter reviews the development and use of interstate compacts for the disposal of low-level radioactive waste (LLRW), and discusses the underlying political dynamics. The central topic matter is the Low-Level Radioactive Waste Policy Act as recently amended by Congress, which authorized the formation of regional state compacts for disposal of our nation's LLRW.

BACKGROUND

Low-level radioactive waste is a relatively benign form of radioactive residue. Low-level waste consists of the ordinary objects that become contaminated through the normal application of nuclear technology. Examples of LLRW include obsolete equipment, worn-out tools, cleaning materials, gloves and other protective clothing, and material used to make smoke detectors, syringes, and even luminous watch dials.

Low-level radioactive waste should be sharply distinguished from the more hazardous high-level radioactive waste. High-level radioactive waste (HLRW) consists either of the unreprocessed spent fuel rods from commercial, research, and defense production reactors or the highly radioactive liquid that results from the reprocessing of the spent nuclear fuel. The public tendency to confuse the less hazardous LLRW with the extremely hazardous high-level variety has hampered efforts to develop an appropriate disposal strategy for LLRW.

Low-level radioactive waste is generated through a variety of common, everyday activities. About half of the nation's LLRW is generated from the operation of the nation's commercial nuclear reactors. Industrial and

Low-Level Radioactive Waste Regulation: Science, Politics, and Fear, Michael E. Burns, Ed., © 1988 Lewis Publishers, Inc., Chelsea, Michigan—Printed in USA.

medical radioactive by-products generated by the nation's factories and hospitals make up the other half. Hospitals generate LLRW by using radioactive materials in diagnostic and therapeutic treatments. Industry generates LLRW through the manufacture of radiopharmaceuticals, watch dials, radiography instrumentation, and a host of other products.[1,2]

LLRW does not pose a substantial or lasting environmental threat to the biosphere. The relatively low levels of radioactivity inherent in LLRW require only marginal shielding and will decay rapidly compared to high-level radioactive wastes. A 1981 Nuclear Regulatory Commission (NRC) Draft Environmental Impact Statement concluded that over long periods (up to a thousand years) the maximum permissible releases of radiation resulting from shallow burial of LLRW would be extremely low and the concomitant threat to the health and safety of the public would be trivial.[3] Because of the minimal public health risk, only shallow burial at closely monitored sites is required for the safe disposal of low-level radioactive wastes.

In the early 1970s, there were six operating LLRW disposal sites scattered across the country, but three of these sites were closed between 1975 and 1979. The three remaining LLRW sites are the Barnwell site in South Carolina, the Beatty site in Nevada, and the Hanford site in south central Washington. Thus, the nation's entire LLRW disposal capacity is limited to these three sites, two of which are located in the West, far from where the vast majority of the LLRW is generated (the East Coast, Southeast, and Midwest).

Reliance on only three LLRW sites proved to be unworkable and wholly unacceptable to the three host states. South Carolina, Nevada, and Washington understandably believed that they were responsible for a disproportionate and inequitable share of the nation's low-level waste disposal burden. In 1979, the governors of Nevada and Washington dramatized their discontent when they closed their respective sites because of some minor transportation and packaging violations. South Carolina also indicated that it would reduce the amount of low-level waste it would accept.[4] All three states were concerned about the political consequences of becoming the nation's nuclear dumping grounds.

The public opposition to out-of-state waste shipments was particularly pronounced in Washington. In 1980, the state's voters overwhelmingly approved Initiative 383, which would have prohibited importation of nonmedical radioactive waste after July 1, 1981, from states not party to a regional interstate compact with Washington. The initiative was challenged in federal court, where it was declared unconstitutional because it violated both the supremacy and commerce clauses of the United States Constitution.[5,6]

THE ORIGINAL FEDERAL ACT

The temporary closing of two of the country's three LLRW disposal sites and the attempt by the state of Washington to prohibit out-of-state waste shipments focused national attention on low-level radioactive waste. Congress quickly began looking for answers. Spurred by the equity complaints of the three LLRW states, the Ninety-sixth Congress passed the Low-Level Radioactive Waste Policy Act (LLRWPA) in December 1980.[7]

As a fundamental federal policy, the act establishes that each state is responsible for providing disposal capacity for all low-level radioactive waste generated within the state. To implement this policy objective, the act encouraged the formation of regional state compacts whenever "necessary to provide for the establishment and operation of regional disposal facilities for low-level radioactive waste." Congress must grant its consent to such compacts before they become operative and enforceable. Further, the LLRWPA exempts from the compact process low-level waste generated from national defense and through federal research and development activities.

The act established an exclusion date of January 1, 1986. The LLRWPA envisioned that after this threshold date, any state party to an approved LLRW regional state compact could refuse to accept wastes shipped from noncompact states for disposal at the compact's designated regional disposal site. The 1986 deadline was apparently set to encourage states to enter into compacts promptly or face the prospects of either having to develop their own low-level disposal site or curtail altogether the production of LLRW.

It is difficult to piece together the congressional intent underlying the Low-Level Radioactive Waste Policy Act. The act was passed in the waning hours of the Ninety-sixth Congress only after it became clear that a significantly more comprehensive bill which also dealt with high-level radioactive waste would not be approved by both houses of Congress. When the conference committee negotiations broke down on the HLRW provisions, Congress extracted the noncontroversial provisions relevant to LLRW from the omnibus legislation. The LLRW provisions were subsequently passed with little formal consideration, leaving a relatively meager legislative history directed specifically at the LLRWPA.

The scant legislative history that does exist suggests that the act's overriding rationale was the need to create more disposal sites on a regional basis. The committee reports emphasized that South Carolina, Nevada, and Washington were reluctant to continue accepting waste unless "national progress" was made "in creating new sites, and in more evenly distributing the burden of the low-level waste disposal."[8] Substantial concern was expressed that if LLRW sites were closed down, essential medical and industrial activities that generate LLRW would be forced to cease.

The legislative history, moreover, suggests a strong preference for a regional approach. One of the House committee reports clearly states that "low-level radioactive waste can more effectively and efficiently be managed on a regional basis."[9] Another House report also determined that the primary responsibility for low-level waste burial lies with the states, because they are "better capable of the planning and monitoring functions relevant to low-level waste."[8]

Compacts

Shortly after passage of this act, states began meeting to discuss the formation of regional compacts for the disposal of LLRW. Six main geographic regions evolved from these initial discussions, with each region consisting of five to eleven states. Two important states, California and Texas, elected at first to pursue independent courses and did not negotiate with any of the compact regions. By the end of 1985, 42 states had ratified at least one regional compact agreement.

Despite the states' best efforts to enter into compacts, no compact ever became fully operative during the lifetime of the original 1980 act. Without congressional ratification, the compacts were without the force of law. Although seven regional compacts, covering 37 states, were introduced into the Ninety-ninth Congress for ratification, none was formally ratified until after the original 1986 exclusionary date.

The fundamental reason for the delay in developing and ratifying regional compacts reflects the politics of radioactive waste. There is an obvious political disincentive peculiar to the selection of nuclear waste disposal sites. Even though the initial act established a 1986 exclusion date, most states realized that this deadline would not be met. As long as three sites remained in operation, many states were reluctant to engage in the political battle of deciding where new disposal sites should be built. The adage "not in my backyard" (often referred to as the "NIMBY syndrome") dominated the political decision making process.[10-12]

Regions that already had operating disposal sites—the Northwest (Hanford), the Southeast (Barnwell), and the Rocky Mountain region (Beatty)—wanted to consummate enforceable compacts as soon as possible. Once these sited compacts were ratified by Congress, their respective regions could exercise the 1986 exclusionary date authority and exclude shipments of low-level radioactive waste generated outside their region. Other parts of the country without available sites (e.g., the Northeast and Midwest) were reluctant to develop compacts because they faced the political hurdle of selecting new LLRW burial sites in their respective states. These nonsited regions obviously preferred to use the existing disposal capacity at Hanford, Barnwell, or Beatty.

The nonsited states hoped the status quo would prevail despite the passage of the 1986 exclusion date. The General Accounting Office (GAO) had already concluded that new disposal sites for LLRW would not be ready in most regions by 1986.[2] Without new disposal sites, the three existing burial sites in Washington, South Carolina, and Nevada would be forced to remain open. As long as states knew that LLRW disposal sites would be available after 1986, there was little incentive for them to organize and approve compacts. In fact, the more states that failed to develop compacts, the more likely it was that the currently operating sites would remain open to take waste. This only enhanced the incentive for some states, acting almost as conscious conspirators, to delay the establishment of regional compacts.

The politics prevailing in Congress also worked against swift ratification of those compacts that were submitted to Congress for approval. The Congress has never expeditiously handled the politically sticky issue of nuclear waste. The political reality of LLRW compacts was that members of Congress from states without disposal sites constituted a congressional majority. Those congressmen from nonsited regions were not about to consent to any compact of sited states that could result in the closure of existing disposal sites.[6]

The 1986 exclusionary date, intended to compel the Congress to ratify compacts before 1986, actually had the opposite effect. It resulted in a political quagmire for Congress and prevented the ratification of any compacts that envisioned closing available sites after 1986. Congressmen from Massachusetts, for example, opposed ratification of a compact for the Northwest because that ratification would effectively close the Hanford site to Massachusetts generators of waste. Parochial politics essentially prevented Congress from ratifying compacts before the 1986 exclusionary date.

Amendments to the Act

These same parochial politics were the driving force behind a series of accommodations that Congress fashioned late in 1985. After intense and intricate compromising, these accommodations manifested themselves as amendments to the 1980 Low-Level Radioactive Waste Policy Act. The result was the Low-Level Radioactive Waste Policy Amendments Act of

1985 (the "new act," "amendments," or "LLRWPAA"), signed into law by the president on January 15, 1986.[13]

In enacting this new law, Congress effectively "cut a deal" that gave both competing parties what they needed. Those who were relying on three existing sites for disposal of their LLRW got a seven-year transitional period, during which waste could still be shipped to Hanford, Barnwell, or Beatty. Those who wanted to close existing sites to out-of-region generators as soon as possible bargained for guarantees that non-sited regions would develop their disposal facilities.

The specific elements of the bargain struck that are incorporated into the new act include the following:

Definitions. As the states began developing compacts in the early 1980s, uncertainties arose concerning the definition of low-level radioactive waste. Some of the regional compacts did not define low-level waste in a fashion consistent with federal definitions and regulations. Efforts to achieve consistency were complicated by changes to the federal definitions as the compacts were being organized.

The act attempts to mitigate the confusion with a consistent federal definition of LLRW. The definition simply states that low-level radioactive waste is any radioactive material that is not high-level waste, spent nuclear fuel, or by-product nuclear material and whatever NRC classifies as LLRW.[14] This certain, uniform federal definition helps resolve uncertainties. However, some confusion will always exist so long as the definition defines LLRW by what it is not.

Division of Responsibility. The newly enacted amendments clearly delineate who is responsible for the disposal of certain types of low-level radioactive waste. The division of responsibility essentially depends on who generated the waste. If the LLRW was generated by commercial firms, state entities, or federal agencies pursuing civilian objectives, the states are responsible; but if the waste was generated by federal agencies to support national security missions, then the federal government is liable for disposal.

The act specifically mandates that individual states are responsible for the permanent disposal of Class A, B, and C LLRW generated within their boundaries.[15] For these waste types, the amendments maintain the federal policy that they can be most safely and effectively managed on a regional basis with interstate compacts.[16]

The federal government is responsible for waste owned or generated by the Department of Energy, by the U.S. Navy as a result of vessel decommissioning, or as a result of any research and development related to atomic weapons production. The states are responsible for any other

LLRW generated by the federal government, such as wastes produced by the National Institutes of Heath. But the federal government is responsible for disposal of any low-level radioactive waste with concentrations of radionuclides in excess of NRC Class C limits.[15]

Interim Period. The most important feature of the new act is the continued availability of existing regional disposal sites for an additional seven years. The 1980 statute allowed LLRW disposal sites to refuse out-of-region wastes starting in 1986. However, the new amendments mandate that the three existing sites remain open to out-of-region wastes for the next seven years, from January 1, 1986, through December 31, 1992.[17]

This seven-year interim access period, however, is subject to several major limitations. The most important limitation is a minimum statutory amount of disposal capacity. The maximum total amount of LLRW that must be disposed of at existing sites during the seven-year interim period is 19.6 million cubic feet (mcf). This aggregate amount breaks down to 2.8 mcf per year. The new statute provides a guidance allocation of this yearly volume cap between the three existing disposal sites. The site in Hanford, Washington, should take about 1.4 mcf per year; the site in Barnwell, South Carolina, about 1.2 mcf per year; and the site in Beatty, Nevada, about 0/0.2 mcf per year.[18]

The new amendments also contain a complex formula for calculating the amount of disposal capacity specifically available to electric utilities for their nuclear power plants. The total volume limit for utility low-level waste over the seven-year interim period is 11.9 mcf, or about 60% of the available disposal capacity for all LLRW. Congressional estimates assume that currently operating and new reactors would need about 11.1 mcf of disposal capacity. The remaining .8 mcf of capacity is part of a "utility set-aside" to be managed by the Department of Energy to cover larger, unexpected volume needs.

Utilities receive precise allocations for each nuclear power reactor in operation, and they may aggregate allocations for more than one reactor in a utility system. Factors considered in determining the appropriate allocations include whether the reactor is a boiling-water or pressurized-water unit, whether the reactor is inside or outside one of the compact regions with operating sites, and whether the first month of operation is during the first four years or subsequent three years of the seven-year interim access period. (The former of each of these criteria means larger allocations for the reactor unit.)[19]

Finally, there is authority for emergency access. During the seven-year interim period (and after), the NRC can require access for LLRW if the lack of disposal capacity results in an "immediate and serious threat to the public health and safety or the common defense and security."[20] The grant of emergency access is available only on the specific request of a generator or state. The access is also temporary; it can last no more than six months and can be renewed only once.[21]

Conditions of Access. The new act creates a series of conditions for access during the seven-year interim period—milestones that states without disposal sites must abide by if they wish to continue to rely upon existing disposal sites for their LLRW.[22] These milestones (and other incentives) were instrumental in gaining the support of the sited states to allow for the interim access period.

Each milestone represents a discrete step in the process of developing new disposal sites for each state or compact region to relieve the burden on the three operating disposal sites. Failure of states or compact regions to meet specific milestones results in denial of access to operating sites for LLRW generators within the state. In some cases, surcharge penalties are imposed during a grace period before access is terminated. The major milestones for new site development and accompanying sanctions are summarized in Table 1.

If the new disposal capacity is not available for nonsited states by 1993, the state must pay the generators of LLRW part of the surcharge money received in rebates. (See description of surcharges and rebates below). If the new disposal capacity is not provided by 1996, the state must take title to and possession of all low-level radioactive waste in the state. This important provision appears to be a backup requirement, aimed at ensuring that nonsited states are ultimately responsible for their own wastes. The state is also obligated to pay any damages resulting from failure to provide a new disposal site.[23]

Surcharges. The authority provided in the new act to levy surcharges is intended to provide additional incentives to drive nonsited states to develop new LLRW disposal capacity. The three states with operating sites are permitted to impose surcharges in addition to regular fees on low-level radioactive waste generated outside their compact region. The surcharges would be charged to the waste generators, even though they are not to blame for the lack of new disposal capacity.

The surcharges escalate during the seven-year interim access period, so that it is less costly for nonsited regions to develop their new disposal sites as soon as possible. The maximum surcharge that can be applied during the first two years (1986–1987) is $10/ft^3 of LLRW; it would increase to $20 the next two years (1988–1989) and double to $40 the final three years (1990–1992).[24]

Rebates are also mandated when states comply with each milestone for new site development. Each time a specific milestone is satisfied, states

Table 1. Major Milestones in the LLRWPA Amendments of 1985

Deadline	Task to be Completed	Penalty for Noncompliance
7/1/86	State must join compact or indicate intent through act of legislature of governor to develop own site.	First 6 months: double surcharge. After 6 months: access denied.
1/1/88	Complete detailed plan for siting of new disposal facility.	First 6 months: double surcharge. Second 6 months: 4 times standard surcharge. After 12 months: access denied.
1/1/90	NRC license application for new disposal facility or governor certifies that waste will be taken care of by 1993.	Access denied.
1/1/92	License application filed for disposal facility.	Triple surcharges.
1/1/93	Disposal provided.	Access denied.

or the relevant compact region will receive a 25% rebate of the amount of money generators from that state paid to the sited region in surcharges. The rebate will be taken from a separate escrow account and it can be used only for specified site development activities.[25] The rebates are obviously intended to act as an incentive by rewarding states for meeting the established milestones.

The enactment of this unusual combination of clarifications, incentives, sanctions, and, above all, the interim seven-year access period laid the foundation for ratification of seven regional compacts. Without all these amendments to the 1980 statute, none of the compacts would have received the formal consent of Congress. The seven compacts ratified include the regional grouping of states from the Northwest, the Central States, the Southeast, the Central Midwest, the Midwest, the Rocky Mountains, and the Northeast. (Table 2 lists the ratified compacts and the 37 states covered.) These seven compacts were ratified by Congress word-for-word as they were enacted by the states representing their respective regions.[26]

There were, however, some very important conditions to the consent of these regional compacts for the disposal of LLRW. The essential condition is that each compact must comply with all of the provisions of the new 1985 amendments to the Low-Level Radioactive Waste Policy Act.[27] This means that each ratified compact must meet all the dictates of the new law, including the milestones. In addition, the effect of the conditional consent is to nullify all those state compact provisions that may come in conflict with federal law or regulations.[28]

THE POLITICS OF ENACTMENT

The mere passage of the 1985 amendments to the LLRWPA through five committees, several subcommittees, both houses of Congress, and finally through the president proved to be a formidable task. The politics inherent to radioactive waste disposal also contributed to the difficulty in passing this legislation. Even Senator Strom Thurmond of South Carolina, a 30-year veteran of the Senate, announced on the Senate floor just before final passage,

> . . . I do not know of any piece of legislation that has required as much skill, as much patience, and as much effort and cooperative effort to pass since I have been in the Senate as this particular piece of legislation . . . [29]

This amazing statement, while perhaps somewhat of an overstatement, does underscore the tricky political compromise that made this legislation acceptable to a majority of the Congress. The original 1986 exclusionary date was a driving force that brought the competing interests — the sited and nonsited states — to the table to develop a compromise. As Congressman Morris Udall of Arizona, a major architect of all nuclear waste legislation, indicated in his usual candid and witty fashion on the House floor,

> Last year, as the . . . "drop-dead" date drew near, it was clear that some compromise between the sited and nonsited states would have to be reached, or the regional compact approach would have died in a noisy clash of parochial interests.[30]

The political compromise crafted proved absolutely essential to final enactment. In exchange for ratification of their compacts, the three sited states agreed to accept a limited amount of LLRW over a seven-year

Table 2. Low-Level Radioactive Waste Compacts and States That
Were Ratified by P. L. 99–240

Northwest
Alaska, Hawaii, Idaho, Montana, Oregon,
Utah, and Washington

Southeast
Alabama, Florida, Georgia, Mississippi,
North Carolina, South Carolina, Tennessee, and Virginia

Rocky Mountains
Colorado, Nevada, New Mexico, and Wyoming

Central
Arkansas, Kansas, Louisiana, Nebraska, and Oklahoma

Midwest
Indiana, Iowa, Michigan, Minnesota, Missouri
Ohio, and Wisconsin

Central Midwest
Illinois and Kentucky

Northeast
Connecticut, New Jersey, Delaware, and Maryland

Note: Three of these compacts (Northwest, Southeast, and Rocky Mountain) already
have available disposal sites. The other four compacts have not yet developed
acceptable LLRW disposal sites.

period. In addition, to ensure that nonsited states made progress in
developing new disposal sites over the seven-year period, the final legislation established milestones, sanctions, and other incentives.[31] The sited
states were not about to be burned again by other nonsited states that
failed, for whatever reason, to develop new disposal sites.

While the basic parameters of the compromise seemed reasonably
straightforward, the final legislative product is incredibly complex. The
detail and specificity used to describe the utility waste allocations, for
instance, confuses even the most knowledgeable expert. Congressman
Manuel Lujan, in his remarks on the House floor, even said that "the
only living human being who completely understands exactly" how the
new public law will operate is the staff person who helped develop the
bill.[32]

This remark is a sad commentary on the way Congress makes public policy. It is certainly disappointing that the only individual who fully understands this legislation is a staff aide who was never elected and who did not cast a vote on passage of the bill. Congress enacted this law without having a thorough appreciation of its specific statutory elements. Much like the 1980 act, the report language, which is intended to clarify uncertainties and resolve ambiguities, is not clear. The House filed two committee reports,[33] but an accompanying Senate report was never developed. The lack of a Senate report is an important omission, since key provisions of the final legislation were Senate provisions — such as the tougher milestones and the mandate that states take title to low-level waste generated in their own states after 1996.

The congressional process in which this legislation was passed was haphazard at best. The Congress was racing both the exclusionary date (January 1, 1986) and the adjournment date for the first session of the Ninety-ninth Congress. The legislation finally passed on December 19, 1985, literally hours before an anxious Congress adjourned for the holiday season. Despite the days of hearings and committee consideration, the final key decisions regarding passage were made in the last hectic and waning hours of the first session. The final passage of the 1985 amendments closely resembled the passage of the original 1980 statute that was criticized in its own right for its quick consideration, bereft of much deliberative thought.[34]

In the haste to finish consideration of the 1985 amendments, some critical issues were left behind. A good example is the appropriate regulatory treatment of mixed waste — those wastes that are regulated both as hazardous and as radioactive wastes. Substantial confusion has emerged as to whether EPA, NRC, or both EPA and NRC regulations should govern the disposal of mixed wastes. Efforts were made to settle this issue legislatively, but it was left unresolved, because time simply ran out. Though some controversy remained on the mixed waste issue, key House and Senate leaders decided it was better to remove the mixed waste provision from the final bill so that some bill could pass before the deadline of January 1, 1986.[35]

While this may have been a politically smart decision, the mixed waste issue remains an uncertainty that could hamper development of new LLRW disposal sites. Congress is currently reviewing the mixed waste issue, but most agree that separate legislation is unlikely. Without specific statutory clarification, the regulatory uncertainty surrounding mixed waste could undermine efforts of some regions in licensing LLRW disposal sites.

This mixed waste issue underscores the failure of Congress to fashion appropriate legislation to address the low-level radioactive disposal issue. It is seemingly impossible to draft perfect legislative remedies in the highly emotional and political environment that dominates nuclear waste issues. The regional parochialism and competition between sited and nonsited regions only aggravates the confusion. One cannot expect the

final legislative product that emerges from this highly charged political process, dominated by deadlines and narrow self-interest, to be a satisfactory or
final resolution of the problem.

PROSPECTS

Given the dismal failure of the original 1980 act and the haphazard
process in which the 1985 amendments were passed, one must wonder if
Congress won't find itself, perhaps when the interim access period ends in
late 1992, trying to legislate another "solution" to the low-level radioactive
waste disposal "crisis." Many thought that the 1980 act, with its regional
approach and its "reasonable" 1986 exclusionary date, would settle the
equity complaints of Washington, South Carolina, and Nevada and result in
more disposal sites. But a series of political and logistical factors doomed the
1980 act and necessitated the 1985 amendments. Has anything changed to
prevent another failure?

One can postulate that three features of the recent amendments reduce the
possibility of another failure. First, the new statute ratifies seven compacts
that represent most of the states and major generators of LLRW.[26] Congressional consent to the seven regional compacts renders them enforceable law
on the same plane with federal law. Despite the lofty objectives of the
original 1980 act to instill equity and to allow for exclusion of out-of-region
waste, none of it was enforceable until individual compacts were ratified by
Congress. The congressional ratification of seven specific regional compacts
in the 1985 amendments to the original act is the first step in securing some
respectability for the regional approach to LLRW disposal. The consent of
Congress makes the compact operative law that can now be enforced by the
three sited regions.

The second feature of the new law that bodes well for success of the
regional disposal concept is the enforcement incentives, such as the milestones, and surcharges.[22,24] The statutory milestones provide a road map of
specific steps nonsited regions must meet in order to continue their use of the
current disposal facilities. The surcharges also make it financially advantageous for the nonsited regions to satisfy those milestones, and site and
license new disposal facilities as soon as possible. The combination of
enforceable milestones and surcharges was clearly intended to prod the
states without disposal sites to develop them. Even considering the grace
periods, any state that failed to meet the milestones would clearly be facing
the distinct possibility of losing access to the current site for disposal of the
LLRW generated within its boundaries.

The last element of the new 1985 amendments that drives the regions
toward successful resolution of the LLRW disposal crisis is the backup

requirement that states take title to the waste.[23] This provision, agreed to during the hectic last hours of congressional deliberation on the statute, seems to mandate that states without adequate disposal facilities take title to the LLRW and assume liability for whatever damages may ensue. While clear legislative history is not available, this provision seems to be warning nonsited regions that if they fail to develop sites, they will still be fully responsible for the waste products. This backup section effectively prevents nonsited regions from ever escaping from their ultimate obligation to deal with their own low-level radioactive waste. The kind of residuary provision forcing the nonsited states to be responsible in the end for their waste certainly reduces the incentives these states have in avoiding the development of new disposal sites.

These three new features of the 1985 amendments, working in tandem, greatly improve the chances of creating a workable and enforceable regional regime for LLRW disposal. The new statute, however, does not totally resolve all uncertainties. Disposal of radioactive wastes (both low-level and high-level) remains a politically explosive issue that may be simply immune to legislative redress—no matter how thoughtful or systematic.[10-12]

The political disincentives to developing new LLRW disposal sites remain. Few governors or state legislative officials care to be party to the politically messy task of deciding where a new LLRW disposal facility should be sited. Even if that initial siting decision is made, a plethora of state and federal regulatory hurdles stand in the way—any of which could easily engender a law suit from local residents trying to keep the site out of their "backyard." The unresolved issue of mixed waste could also complicate licensing of new sites. Some nonsited states may decide that the price for noncompliance with the new statute, measured in higher surcharges, is entirely acceptable compared to the political repercussions of siting LLRW sites.

The mandated milestones and the backup requirement that the states take title to LLRW, however, make it nearly impossible for nonsited states to ignore the new act. Nonetheless, one can easily envision a scenario of several states developing ostensibly legitimate problems in siting new disposal facilities.

These states, facing deadlines and surcharge penalties, petition their congressmen for help. Sympathetic members of Congress champion efforts to weaken the new statute by relaxing the milestones and penalties, or, for whatever reason, simply exempting certain regions. One congressional exemption leads to another, and the complete unraveling of the statute and the regional approach to LLRW disposal is not far behind. Another possibility, perhaps now occurring in North Carolina, is that a central state in a regional compact may decide to withdraw, leaving the compact without a host state.

These scenarios may seem a little farfetched, but they do suggest that the new 1985 amendments are not a panacea. As long as the political difficulties associated with developing new disposal sites outweigh the enforcement disincentives of the new statute, there is always the distinct possibility of future congressional or state actions to weaken that law. Subsequent amendments to delay the law's milestones and relax the enforcement penalties would not be the first time legislative bodies have undone a statutory compromise.

On balance, the 1985 amendments to the Low-Level Radioactive Waste Policy Act greatly enhance the prospects of developing an enforceable and workable regional disposal regime for LLRW. The combination of ratified compacts, milestones, surcharges, and other enforcement tools drives nonsited states toward developing new LLRW disposal facilities. However, the special political impediments unique to radioactive waste always create uncertainties. The overriding political dynamics of radioactive waste disposal—and not the federal act—may continue to govern the ultimate resolution of this dispute.

This is why most of the observers who closely followed the enactment of the Low-Level Radioactive Waste Policy Act Amendments would not be surprised if a new disposal crisis emerged in about 1992. One can certainly speculate that the new federal law merely postponed the political decisions necessary to resolve how our country should dispose of its low-level radioactive waste.

REFERENCES

1. American Nuclear Society, "Low-Level Radioactive Waste" (Chicago, IL: American Nuclear Society [PPS-11], August 1982).
2. U.S. General Accounting Office, "Regional Low-Level Radioactive Waste Disposal—Progress Being Made but New Sites Will Probably Not Be Ready by 1986" (Washington, DC: U.S. General Accounting Office [GAO/RCED-83-48], April 11, 1983), pp. 2-3.
3. U.S. Nuclear Regulatory Commission, "Draft Environmental Impact Statement on the Proposed Rule 10 CFR Part 61, Licensing Requirements for Land Disposal of Radioactive Waste" (Washington, DC: U.S. Nuclear Regulatory Commission, September 1981), p. 64.
4. Melson, G., "Time to Take Control: The States and Low-Level Radioactive Waste," *State Legislatures*, 7(6):7-11 (1981).
5. *Washington State Bldg. & Const. Trades Council* v. *Spellman*, 518 F. Supp. 928 (E.D. Wash., 1981), aff'd 684 F.2d 627 (9th Cir. 1982), cert. denied, 461 U.S. 913 (1983).

6. Peckinpaugh, T. L., "An Analysis of Regional Interstate Compacts for the Disposal of Low-level Radioactive Wastes," *J. Energy Law & Policy* 5(1): 41–44 (1983).

7. Low-Level Radioactive Waste Policy Act, P.L. 96–573 (LLRWPA), 94 Stat. 3347 (codified in part at 42 U.S.C. §2201 [Supp. V,1981]).

8. "Committee Report to Accompany the Atomic Energy Act Amendment of 1980," Committee on Interior and Insular Affairs, House Report 1382, Part 2, Ninety-sixth Congress, Second Session (1980), p. 25.

9. "Committee Report to Accompany the Nuclear Waste Policy Act," Committee On Interstate and Foreign Commerce, House Report 1382, Part 1, Ninety-sixth Congress, Second Session (1980), p. 34.

10. Omag, J., "States are Juggling A-Waste Disposal Like Hot, Ah, Potato," *Wash. Post*, March 2, 1983, p. 3.

11. Buckser, A. S., "Environmental: Compacting Radioactive Waste," *Harvard Polit. Rev.* 10(19) (Summer 1983).

12. Stanfield, R. L., "Radioactive Waste Can't Find a Home," *National Journal*, January 4, 1986, pp. 38–39.

13. P.L. 99–240 (H.R. 1083), January 15, 1986, 99 Stat. 1842.

14. LLRWPA (42 U.S.C. 2021b et seq.) as amended by P.L. 99–240, §2(a).

15. LLRWPA, §3.

16. LLRWPA, §4(a).

17. LLRWPA, §5(a).

18. LLRWPA, §5(b).

19. LLRWPA, §5(c). A legislative summary developed by the American Nuclear Energy Council neatly summarizes the features of the formula for determining the utility allocations. See "Summary of the Low-Level Radioactive Waste Policy Amendments Act of 1985 (P.L. 99–240)" (Washington, DC: American Nuclear Energy Council, January 23, 1986).

20. LLRWPA, §6(a).

21. LLRWPA, §6(b)-(d).

22. LLRWPA, §5(e).

23. LLRWPA, §5(d)(2)(C).

24. LLRWPA, §5(d)(1).

25. LLRWPA, §5(d)(2)(A) and §5(d)(2)(F).

26. P.L. 99–240, Title 2, Subtitle B.

27. P.L. 99–240, §212.

28. LLRWPA, §4(b)(3) and §4(b)(4).

29. *Congressional Record*, December 19, 1985, p. S18117. Senator Thurmond was an important and persistent force in the final passage of the 1985 amendments. Indeed, Senator Simpson described Thurmond's persistence in colorful terms: "Thanks to Strom Thurmond for his bulldog persistence, best described as, I believe, if we could say so, a Brahma bull on low-level waste matters" (*Congressional Record*, December 19, 1985, p. S18123).

30. *Congressional Record*, December 9, 1985, p. H11410.

31. Lujan, M., *Congressional Record*, December 9, 1985, p. H11410.

32. *Congressional Record*, December 9, 1985, p. H11411.

33. Interior and Insular Affairs Committee Report, House Report 99-314, Part 1, Ninety-ninth Congress, to accompany H.R. 1083, October 22, 1985; and Energy and Commerce Committee Report, House Report 99-314, Part 2, Ninety-ninth Congress, to accompany H.R. 1083, December 4, 1985.

34. Steel, J. B., and Bartlett, D. L., "How the Act Made It through Congress: Lots of Self-interest and Little Thought," special reprint of *The Philadelphia Inquirer*, November 13-20, 1983, p. 45.

35. Lujan, M., *Congressional Record*, December 19, 1985, p. H13076.

CHAPTER **3**

Interstate Compacts for Low-Level Radioactive Waste Disposal: A Mechanism for Excluding Out-of-State Waste

Robert L. Glicksman

INTRODUCTION

At the end of World War II, the United States extended its atomic power capabilities from military to civilian uses. The federal government has since retained control over those aspects of handling nuclear materials that endanger the public health, the public safety, and the environment. Some of the most serious environmental threats have stemmed from the absence of a safe and effective disposal mechanism for radioactive waste. The federal government assumed the responsibility for locating and constructing high-level radioactive waste disposal sites when Congress enacted the Nuclear Waste Policy Act of 1982.[1] It has been less eager to take charge of low-level radioactive waste. By the late 1970s, South Carolina, Nevada, and Washington were the only three states with operating low-level waste disposal facilities. Officials in these states complained that the failure of Congress to assume responsibility for low-level wastes unfairly burdened them with the wastes generated throughout the nation.

Congress responded by passing the Low-Level Radioactive Waste Policy Act of 1980 (LLRWPA),[2] which attempted to remove these burdens by providing for safe, efficient disposal of low-level radioactive waste on a regional basis.[3] This chapter will discuss the allocation of state and federal responsibility for achieving that objective under the act as amended in 1985. The chapter first discusses two provisions of the United States Constitution that in general limit state power to regulate nuclear materials, and in particular prohibit state attempts to ban the

Low-Level Radioactive Waste Regulation: Science, Politics, and Fear, Michael E. Burns, Ed., © 1988 Lewis Publishers, Inc., Chelsea, Michigan – Printed in USA.

import and disposal of out-of-state wastes. These two constitutional provisions are the Supremacy Clause[4] and the Commerce Clause.[5]

The chapter next focuses on the mechanism chosen by Congress to spread the burden of low-level radioactive waste disposal: the formation of interstate compacts for constructing and operating disposal sites. It then examines the impact of a third constitutional provision, the Compact Clause, on state efforts to reject out-of-state wastes. This provision permits the states to enter into congressionally approved interstate compacts or agreements. The chapter analyzes the extent to which an interstate compact is an enforceable "contract," the legal rights and obligations of states participating in such agreements, and the thorny problems that may arise from trying to use interstate compacts to solve the low-level waste disposal problem.

STATE EXCLUSIONARY POWER UNDER
THE LOW-LEVEL RADIOACTIVE WASTE POLICY ACT

The LLRWPA, described more fully in Chapter 2 of this book, makes each state, either alone or in cooperation with other states, responsible for disposing of low-level radioactive waste generated within its borders.[6] Because Congress concluded that low-level waste can be most safely and effectively managed on a regional basis,[7] the LLRWPA authorizes the states to enter into compacts or agreements for establishing and operating regional low-level waste disposal sites.[8] The statute provides the states with a strong incentive to enter such compacts by giving compact members the right to exclude waste generated outside the compact region from regional disposal facilities.[9] Compacting states have no authority to exclude nonregional waste, however, until Congress has by law consented to the compact.[10] Congress reserves the right to withdraw its consent every five years.[11] Thus, the LLRWPA grants limited authority to the states to bar the in-state disposal of wastes generated elsewhere.

STATE AND FEDERAL POWER TO REGULATE
NUCLEAR WASTE DISPOSAL

Unless a state enters an interstate compact under the LLRWPA, it may not be able to close its borders to waste generated in another state. Two different provisions of the United States Constitution appear to prohibit a state's attempt to isolate itself in this way: the Supremacy Clause and the Commerce Clause.

The Supremacy Clause

The Supremacy Clause provides that the federal Constitution and federal laws made under it are the supreme law of the land.[12] State court judges are bound to apply governing federal law, and state statutes that are inconsistent with federal laws are invalid. If Congress precludes the states from adopting any legislation in a particular area covered by federal law, it is said to "preempt" state law. Congress may preempt state legislative authority in a particular area expressly or by implication.[13] The courts are willing to find implicit preemption where the federal regulatory program is so pervasive that it appears that Congress did not want to permit supplementary state legislation.[14] Implicit preemption may also result where the federal interest in a particular area is so dominant that it precludes enforcement of state laws on the same subject.[15]

Even where Congress does not entirely displace state law by occupying a particular regulatory area, state law is preempted to the extent that it actually conflicts with federal law.[16] Such a conflict exists either where it is physically impossible to comply with both federal and state law, or where state law frustrates congressional purposes.[17]

The Supreme Court concluded in a 1983 decision that when Congress passed the Atomic Energy Act of 1954, it occupied nearly the entire field of nuclear safety concerns.[18] As a result, with a few limited exceptions, no state may pass legislation to protect its citizens from the hazards of nuclear materials, although to achieve other goals, such as protecting the state's economy, it may regulate certain aspects of their generation and use.[19] Based on this federal occupation of the field of nuclear safety, a federal court of appeals in 1982 invalidated an Illinois law prohibiting the storage or disposal in Illinois of spent nuclear fuel used in any power-generating facility located outside the state.[20] According to the court, the Atomic Energy Act of 1954 has preempted state authority to regulate the storage, shipment for storage, and disposal of spent nuclear fuel if the state's goal is to protect public health and the environment.[21]

Another federal appellate court in 1983 held unconstitutional the state of Washington's 1980 initiative prohibiting the transportation, storage, and disposal within Washington of low-level radioactive waste produced outside the state.[22] Despite the existence of the LLRWPA, which authorizes states to enter into waste disposal compacts barring wastes generated outside the compact region, the Washington initiative was preempted by the Atomic Energy Act. The court concluded that the low-level waste statute does not give the states absolute authority over low-level waste disposal within their own borders. Rather, the statute augments state regulatory authority only if the state enters into an interstate compact approved by Congress. Although Washington had entered a waste disposal compact with Oregon, Idaho, Montana, and Utah, Congress had not yet approved it at the time of the court's decision.[23] Accordingly, the state's law purporting to close its borders to out-of-state wastes was an invalid intrusion into a field preempted by federal law.

The Commerce Clause

The Commerce Clause gives Congress the power to regulate commerce among the several states.[24] The Supreme Court has concluded that this affirmative grant to the federal legislature has "negative implications" for the states' ability to regulate interstate commerce. Specifically, the Commerce Clause restricts the states from erecting barriers to the free flow of interstate commerce.[25] The Commerce Clause limits state power to regulate the interstate flow of economically useful commodities.[26] It also applies to state control over waste disposal, even though the regulated activity involves traffic in "bads" rather than in "goods," and would normally not be thought of as "commerce":[27]

> The efficient disposal of wastes is as much a part of economic activity as the production that yields the wastes as a by-product, and to impede the interstate movement of those wastes is as inconsistent with the efficient allocation of resources as to impede the interstate movement of the product that yields them.[28]

Despite the "negative implications" of the Commerce Clause, the Constitution does not prohibit all state regulation of interstate commerce. A state law affecting interstate commerce is valid if it regulates evenhandedly, accomplishes a legitimate local purpose, and has only an incidental effect on interstate commerce.[29] When Illinois and Washington tried to ban the import of waste generated elsewhere, the courts of appeals ruled that the Illinois spent-fuel statute and the Washington low-level waste initiative fell outside the scope of legitimate state regulation.[30] The court in the Washington case, for example, said that Washington's Initiative 383 did not regulate evenhandedly—it discriminated against out-of-state waste by banning the in-state disposal of waste generated outside the state, but not of internally generated material.[31] Initiative 383 also lacked a legitimate local purpose. The state based its ban on the disposal of out-of-state waste on the belief that releases of radioactive materials to the environment jeopardize the health and welfare of the state's residents. The state's other actions cast doubt on this explanation, though; Washington permitted low-level wastes generated within the state to be disposed of there, but produced no evidence that internally generated wastes posed a lesser health hazard than externally generated ones.

Finally, Initiative 383 had more than an incidental effect on interstate commerce. The court concluded that "closing Washington's borders would significantly aggravate the national problem of low-level waste disposal" because the state then received 40% of the country's low-level waste and all of the country's absorbed-liquid low-level waste.[32]

Congress may authorize state regulatory activity that would otherwise be prohibited as an impermissible burden upon or improper attempt to discriminate against interstate commerce.[33] The court in the Washington case concluded, however, that Congress had not sanctioned Washington's regulatory efforts when it passed the LLRWPA in 1980. According to the court, that act authorizes a state to ban low-level waste from outside the state only if the state enters an interstate compact and the compact is approved by Congress.[34] Congress had not yet approved the compact of which Washington was a member.

Arguments Supporting a State Embargo

Some legal commentators have argued that the states have broader authority to bar the disposal of out-of-state wastes than the decisions in the Washington and Illinois cases recognized. These writers contend that the LLRWPA ceded to the states all responsibility for regulating low-level radioactive waste, even if the regulation is meant to protect the public health and the environment, and regardless of the degree of burden the regulation imposes on interstate commerce.[35] The argument that Congress delegated to the states all responsibility over civilian low-level nuclear waste, "subject to no express limitation,"[36] however, seems to ignore the LLRWPA's requirement that states obtain congressional consent before implementing compact provisions that exclude waste generated outside the compact region. To permit states to ban out-of-state wastes without a congressionally approved compact would be inconsistent with the apparent desire of Congress to retain supervision over state disposal activities that might have an adverse effect on the national interest.[37]

A state seeking to support the constitutionality of excluding out-of-state low-level waste has several more tenable arguments. The state may contend, for example, that neither the Atomic Energy Act nor the LLRWPA preempts a ban on waste generated outside the state if the ban's purpose is to protect the state's economy by avoiding the need to construct expensive new disposal facilities.[38] If a court accepts such a contention, the Supremacy Clause will not prohibit a state's exclusion of out-of-state wastes. Texas may defend its proposed embargo on out-of-state waste against a Commerce Clause attack on a different ground. The state plans to assert that, by constructing and operating its own disposal site, it is acting as a direct market participant rather than as a regulator of the private activities of others.[39] The Supreme Court has permitted what would otherwise be improper discrimination against out-of-state activities in similar contexts.[40]

The United States Supreme Court may soon address the issue of the scope of state authority to exclude out-of-state low-level waste. In 1985, New Jersey filed a complaint before the Court alleging that Nevada improperly required New Jersey to obtain a permit for transporting radium-contaminated dirt into Nevada. According to New Jersey, the LLRWPA preempts Nevada's attempt to impose permit requirements as a prerequisite for access to the low-level waste disposal site in Beatty, Nevada. New Jersey also alleged that Nevada's requirement for a transportation permit for radium-contaminated dirt imposes an improper burden on interstate commerce.[41] The Supreme Court refused to issue a preliminary injunction against Nevada's attempt to bar the wastes, but it also refused to dismiss New Jersey's complaint.[42]

INTERSTATE COMPACTS FOR LOW-LEVEL RADIOACTIVE WASTE DISPOSAL

The Scope and Application of the Compact Clause of the Constitution

The Supremacy and Commerce Clauses appear to prohibit a state from closing its borders to low-level wastes generated outside the state, unless Congress approves of that kind of state prohibition. The LLRWPA provides approval in limited circumstances; states may refuse to accept out-of-state waste by entering into congressionally approved compacts with other states.[43] This statute is not the first example of the use of interstate compacts to resolve difficult national problems among states whose interests differ from those of the federal government or of one another.[44] The Supreme Court has often encouraged states to use interstate compacts to settle disputes.[45] According to the Court, interstate compacts are "more than a . . . device for dealing with interests confined within a region . . . [They are] also a means of safeguarding the national interest."[46] States have responded to this invitation by forming compacts to settle boundary disputes with other states,[47] to govern the interstate transfer of prisoners,[48] to allocate natural resources to which more than one state has access (such as water in interstate rivers),[49] and to limit the discharge of pollutants by one state into another.[50]

The Constitution limits the states' authority to resolve disputes through interstate compacts. The Compact Clause prohibits any state from entering a compact with another state without congressional consent.[51] Despite the apparently plain language of the Compact Clause, however, the Supreme Court has decided that not all interstate compacts require congressional consent.[52] The Compact Clause was meant to protect the rights and interests of states other than the parties to the agreement.[53] Congressional supervision of interstate compacts safeguards the interests of nonmember states and the nation.[54] These interests are affected only by agreements that tend to increase the political power of the compacting states at the expense of the national interest.[55] The Court has therefore concluded that only these compacts require congressional consent.[56]

The Court has provided several examples of interstate compacts that do not require congressional approval: an agreement by one state to purchase land within its borders owned by another state, an agreement by one state to ship merchandise over a canal owned by another, an agreement to drain a malarial district on an interstate border, and an agreement to combat an immediate threat, such as an epidemic.[57] None of these agreements would concern the national interest and thus require congressional approval. The following agreements, on the other hand, would be covered by the Compact Clause: treaties of alliance for purposes of peace or war; treaties of confederation for purposes of mutual government, political cooperation, and the exercise of political sovereignty; agreements conferring general commercial privileges; and agreements settling interstate boundary disputes.[58]

Despite a recent tendency of the Supreme Court toward narrow definition of the category of interstate compacts subject to congressional consent,[59] low-level waste compacts appear to require such approval. A compact purporting to give its members the right to exclude wastes generated in nonmember states would disrupt the free flow of commerce, as the court of appeals concluded in the Washington Initiative 383 case.[60] Such a compact would enhance state power at the expense of the federal government and would therefore be subject to the requirements of the Compact Clause.

Two prominent scholars on interstate compacts have formulated a slightly different test for deciding whether a compact requires congressional approval: it does if it brings state action into a field of federal-state sensitivity.[61] Compacts for the disposal of low-level radioactive wastes would regulate nuclear materials for health, safety, and environmental reasons. Congress marked this field as an area of almost exclusive federal concern when it passed the Atomic Energy Act of 1954, which largely preempted state activity in this field.[62] The Senate Committee on Energy and Natural Resources recently remarked that neither the

LLRWPA, as amended in 1985, "nor any compact ratified under its terms is intended to affect or alter the federal-state relationship under the Atomic Energy Act."[63] Interstate compacts restricting the disposal of out-of-state low-level waste therefore encroach upon an area of federal-state sensitivity, and are subject to the congressional consent requirement of the Compact Clause.

The Forms and Effects of Congressional Approval

The Forms of Congressional Approval

The Constitution does not specify the form in which Congress must consent to interstate compacts; that is for Congress to decide.[64] Congress need not expressly approve each compact provision.[65] Consent may be implied—for example, if Congress either appears to acquiesce in the executed compact by aiding in its enforcement or appears to ratify the compact in some other way.[66] Nothing less than formal, explicit consent, however, is likely to make a compact for low-level radioactive waste disposal effective. The LLRWPA prohibits any compact restricting the use of regional disposal facilities from taking effect until "the Congress by law consents" to it.[67] Consent "by law" appears to preclude any means of approval less formal than passage of a statute approving the compact.

Congress may place conditions on its consent to an interstate compact.[68] The LLRWPA requires that every compact authorize Congress to withdraw its consent five years after the compact has taken effect.[69] Through this periodic review provision, Congress can monitor the operation of an approved compact to see if it is dealing adequately with the low-level waste problem. The House Committee on Energy and Commerce has asserted that, despite this provision, "there is no clear limit upon the authority of any future Congress to withdraw consent" to previously approved compacts.[70] In other words, the committee believes that Congress can withdraw its consent even before the expiration of five years from the time of compact approval. Two courts, on the other hand, have taken a diametrically opposing position, concluding that Congress may never reserve the right to alter or repeal its consent to an interstate compact.[71] If the courts' view is correct, then the LLRWPA's five-year withdrawal provision may be void, since that provision essentially reserves the right to repeal congressional consent once every five years.

A further issue arises when Congress imposes conditions on a compact member after that state has ratified the compact. Some legal analysts have argued that if the congressional conditions result in significant changes in the rights and obligations of the compact parties, the compact is not effective until the state has reratified it.[72] The 1985 amendments to the LLRWPA create significant new obligations for the compacts covering the existing disposal sites in South Carolina, Washington, and Nevada. The members of these compacts, which were approved by Congress in the same legislation containing the 1985 amendments, were not aware when they ratified the compacts, for example, that the Barnwell, Richland, and Beatty sites would have to accept millions of cubic feet of nonregional low-level waste between 1986 and 1992.[73] At the time the three states ratified the compacts, the LLRWPA authorized compact states to ban all nonregional wastes as of January 1, 1986. Despite the increased obligations resulting from the 1985 amendments, the House Committee on Interior and Insular Affairs nevertheless contends that the three states need not formally reratify. According to the committee, "there are no known instances in which a court has held a compact invalid for lack of state reratification after any amendment or condition has been imposed by Congress."[74] The courts and commentators agree that reratification is not always necessary. State parties, by acting under a compact, may implicitly assume and agree to the conditions placed on them by Congress.[75] Thus, the three states with operating disposal sites may implicitly consent to any changes in their compact obligations resulting from the 1985 amendments by accepting nonregional waste in accordance with those amendments.

The Effects of Congressional Approval

When Congress enacts a law approving an interstate compact, the compact becomes binding as a matter of federal law.[76] An approved interstate compact, like all other federal law, supersedes any conflicting state laws under the Supremacy Clause.[77] Some of the low-level radioactive waste compacts approved by Congress in 1985 make this point expressly. The Midwest Interstate Compact, for example, provides that "for purposes of this compact, all state laws or parts of laws in conflict with this compact are hereby superseded to the extent of the conflict."[78] An additional restraint on state legislative enactments passed after an interstate compact is congressionally approved is derived from the contractual nature of such a compact. The Contract Clause of the Constitution prohibits the states from enacting any law impairing the obligations of contracts.[79] The effect of a compact upon a conflicting federal law or regulation is less clear. One commentator has suggested that an interstate compact, even if congressionally approved, nevertheless is inferior to other federal laws or regulations.[80] If, on the other hand, a compact codified by statute into federal law (such as the seven compacts approved in the 1985 amendments to the LLRWPA) is of equal status with other federal laws, then one might expect the most recently enacted of two conflicting federal laws to prevail. The resolution of such a conflict could have significant implications, since some of the compacts differ with federal statutes or regulations on matters as fundamental as the definition of low-level radioactive waste.[81] Some of the low-level waste compacts approved in 1985 seek to avoid such conflicts. The Southeast compact, for example, provides that nothing in the compact will "abrogate or limit the applicability of any act of Congress or diminish or otherwise impair the jurisdiction of any federal agency," such as the Nuclear Regulatory Commission.[82] To the extent conflicts between compact provisions and other federal laws remain, those other laws will usually prevail. The 1985 amendments to the LLRWPA provide that unless that act expressly provides otherwise, nothing in the LLRWPA or in any compact formed under it limits the applicability of any federal law or diminishes the jurisdiction of any federal agency.[83]

The Compact as a Contract

Contractual Nature

An interstate compact is in some respects simply a contract between the states that join it.[84] The rights of the party states should therefore be governed in part by the usual rules of contract formation. For example, to form a valid contract, each contracting party must furnish what the law calls "consideration."[85] In other words, each party must promise to give up something of value in exchange for the other parties' promise to do likewise.

The Northwest Interstate low-level waste compact contains one of the more unusual exchanges of consideration. The parties to that compact recognize that

> the issue of hazardous chemical waste management is similar in many respects to that of low-level waste management. Therefore, in consideration of the state of Washington's allowing access to its low-level waste disposal facility by generators in other party states, party states such as Oregon and Idaho that host hazardous chemical waste disposal facilities will allow access to [those] facilities by generators within other party states.[86]

Since the Supremacy and Commerce Clauses appear to prohibit Oregon and Idaho from refusing to accept hazardous waste generated in Washington,[87] Oregon and Idaho may have promised in this provision to carry out a task they were already under a preexisting legal duty to perform. If so, then the two states' promise to accept Washington's hazardous waste would not by itself supply the consideration necessary to bind Washington to its reciprocal promise to continue accepting low-level wastes generated in Oregon and Idaho. Both sets of promises would be binding, though, if the compact contained other promises imposing on Oregon and Idaho additional new obligations. Other compacts, for example, obligate each state to take a turn in hosting low-level waste facilities.[88] An agreement to rotate host duties furnishes the consideration necessary for a binding contract.

Enforceability

An interstate compact, however, is more than a simple contract between consenting parties. Because such a compact represents an agreement between sovereign entities, a breach of its provisions creates difficult enforcement problems. The Supreme Court has the final authority to rule on the meaning of compact terms[89] and to enforce the obligations of all party states. The Court's enforcement powers are not unlimited. It may not order relief inconsistent with the express terms of any compact approved by Congress,[90] since such an order would infringe upon the authority of the federal legislative branch to make the law.

Although the Supreme Court may issue an injunction ordering a state to fulfill its obligations to other compact members,[91] it has been reluctant to use such decrees to force breaching states to comply with compact terms.[92] Fortunately, the Court has avoided the practical problem of enforcing such a decree, since states defaulting on compact obligations typically comply voluntarily eventually.[93] The attempts by state governors in the late 1970s to physically bar entry of out-of-state low-level radioactive wastes,[94] though, raise the specter of a confrontation between the federal courts and an intransigent state official. New Jersey's recent complaint before the Supreme Court contends that Nevada officials have publicly stated their determination to exclude New Jersey's radium-contaminated dirt from the Beatty low-level waste site.[95]

The federal government could initiate contempt proceedings against state officials who refused to comply with federal court orders enforcing compact obligations.[96] As a last resort, the federal executive branch could use military force,[97] just as it did to insure adherence to school desegregation decrees in the 1950s.[98] The Supreme Court noted in that context that if the "states may, at will, annul the judgment of the courts of the United States, and destroy the rights acquired under those judgments, the Constitution itself becomes a solemn mockery."[99] Congress may also enforce compact obligations through its legislative power.[100]

The Status of Low-Level Waste Compacts Before Congressional Approval

An interstate compact is not governed solely by common-law contract principles. The Compact Clause's requirement of prior congressional approval makes each compact "a constitutional creation."[101] Even if an interstate compact satisfies all of the normal rules for contract creation, a compact not yet approved by Congress has no legal force against states that are not parties to it or against the federal government.

A more difficult question is whether member states are bound by the compact's provisions from the time of its execution. Since the Compact Clause prohibits a state from entering any interstate compact without congressional consent, one can argue that a compact is unenforceable, even among compact members, until Congress approves it. Under this view, the Compact Clause removes the legal capacity of a state to enter a binding agreement with any other state absent congressional approval. At least one Supreme Court decision, on the other hand, appears to indicate that member states may incur obligations among themselves even before congressional approval,[102] although the Court has never definitively resolved this question. The LLRWPA reflects an intermediate view: states are bound by some, but not all, compact provisions during the interval between compact execution and congressional consent. Although provisions purporting to exclude nonregional waste from regional disposal facilities are not effective until Congress approves the compact,[103] an individual state may be bound by the compact's other provisions as soon as it enacts consent legislation.[104]

At the time it passed the 1985 amendments to the LLRWPA, Congress approved seven compacts representing 37 states.[105] Some of the remaining states are in the process of forming additional compacts, and the configuration of existing compacts may change, since all of the approved compacts include provisions for membership withdrawal and revocation.[106] States joining new compacts can avoid incurring obligations before congressional approval by deferring the effective date of the compacts until such approval. The Southeast interstate compact, for example, protected its members by providing that "the consent of Congress shall be required for the full implementation of this compact. [Certain] provisions . . . shall not become effective until the effective date of the import ban . . . as approved by Congress."[107]

Special Problems of Compact Integration

The LLRWPA envisions a network of regional low-level waste disposal facilities that, by 1993, will spread the burden of handling such waste beyond the three states with currently operational sites. Most of the states have joined one of the seven congressionally approved regional compacts.[108] It is possible, however, that one of the remaining states with large waste generation capabilities will meet with resistance if it tries to join a compact whose members have already agreed to construct a facility and are reluctant to accept large additional volumes of waste there. It is not clear that the LLRWPA provides an effective mechanism for integrating such a state into the network of regional compacts.

It is unlikely that a court would force an established compact to accept as a member a large generator which has unsuccessfully sought to join that compact. Such relief would be inconsistent with the voluntary nature of most contractual undertakings. It might also conflict with the terms of the compact. Most of the approved compacts restrict eligibility for membership.[109] If a state fails to meet these requirements, a court cannot force compact members to accept the excluded state. The courts lack the power to grant relief inconsistent with a congressionally approved interstate compact.[110] Alternatively, an excluded state might seek legislative assistance. If the state has been excluded from a compact that has not yet been congressionally approved, Congress could condition its consent upon acceptance of the excluded state. If a state is not permitted to join an approved compact, Congress might threaten to withdraw its consent at the next five-year interval. Such a threat would be ineffectual if Congress lacked the power to withdraw its consent to a previously approved interstate compact.[111]

A state's refusal to join any compact would create different problems. Several states, including Maine, Massachusetts, New York, and Texas, have considered this "single-state option."[112] These states each plan to construct a disposal site exclusively for wastes generated within the state.[113] Unless the LLRWPA authorizes such states to exclude out-of-state waste, though, the Supremacy and Commerce Clauses probably require them to accept waste generated elsewhere.

Unfortunately, the LLRWPA does not indicate whether noncompact states with their own disposal facilities have the same exclusionary authority as regions with congressionally approved compacts. There is evidence that Congress consciously avoided addressing the question of a noncompact state's exclusionary authority when it enacted the LLRWPA in 1980.[114] The 1985 amendments do little to clarify the issue.[115]

The amended statute clearly authorizes states to establish their own disposal sites, thereby avoiding penalties for failing to meet the milestones for disposal site construction and licensing between 1986 and 1993.[116] Texas has asserted that the Senate debate on the 1980 version of the LLRWPA also indicates Congress's intent to vest single states operating disposal facilities with the power to exclude out-of-state waste.[117] Furthermore, the amended LLRWPA makes each state responsible "for providing, either *by itself* or in cooperation with other states,"[118] for the disposal of low-level waste. A state that sets up a disposal facility exclusively for state generators arguably fulfills its statutory responsibilities[119] and should be granted the same exclusionary authority as states meeting their disposal capacity obligations through compacts.

The arguments against single-state exclusionary authority, however, appear more persuasive. The LLRWPA provides that "[a]ny authority *in a compact* to restrict the use of regional disposal facilities" to wastes generated within the compact region is not effective until Congress approves the compact.[120] The act defines a "compact" as "a compact entered into by two or more states. . . ."[121] Since the statute's only references to exclusionary authority relate to the exclusion of nonregional waste by the members of an approved interstate compact, the statute seems to limit that authority to two or more states which have joined a congressionally approved interstate compact.[122] The legislative history is even more direct. The 1980 House report stated that "[only] states which are a party to the interstate compact would have the authority[to exclude nonregional waste]."[123] One of the 1985 House reports stresses that "states acting alone are not . . . considered . . . to constitute compacts . . . ."[124]

Assuming that the LLRWPA provides exclusionary authority only to compact members, a noncompact state could still attempt to isolate itself from wastes generated elsewhere by agreeing to accept such wastes, but charging disposal fees so high that out-of-state generators would ship their wastes elsewhere. Such a scheme, however, would probably violate the Commerce Clause if preferential access to the facility were afforded to generators within the state.[125] Furthermore, the courts are unlikely to permit a state to achieve indirectly what it is prohibited by the Supremacy and Commerce Clauses from doing directly.[126]

CONCLUSION

Existing legal principles pose a formidable barrier to state attempts to exclude low-level radioactive wastes originating outside the state. Absent congressional authorization, the Supremacy Clause of the United States Constitution appears to prohibit a ban on the disposal of out-of-state low-level radioactive waste, if the purpose of such a ban is to protect the public against health and safety hazards. The federal government has largely preempted the field of nuclear safety regulation. An attempt to exclude out-of-state wastes for economic reasons is more likely to withstand an attack on preemption grounds. A second obstacle to a state's attempt to exclude out-of-state waste is the Commerce Clause, which prohibits state legislation that discriminates against or has more than a merely incidental effect on interstate commerce. Absent congressional authorization, it is difficult to conceive of state legislation designed to bar out-of-state wastes that would clear these hurdles.

The LLRWPA provides only limited authority to exclude out-of-state wastes. The three states with disposal sites already in operation must continue to accept limited amounts of wastes generated throughout the nation until the end of 1992. After that time, access to the sites may be restricted respectively to the members of the Northwest, Southeast, and Rocky Mountain interstate compacts. Congress has approved several other compacts, whose members may exclude nonregional waste as soon as they construct and license a low-level waste disposal facility. Congress apparently did not intend to authorize a noncompacting state to restrict access to its own disposal site to low-level waste generated within the state. Thus, individual states do not seem to have the power to exclude out-of-state low-level wastes, and even the ability of compacting states to do so remains subject to congressional scrutiny and supervision.

ACKNOWLEDGMENT

The author thanks Nancy M. Wilson, University of Kansas School of Law, Class of 1987, for her very helpful research assistance.

REFERENCES

1. P.L. 97–425, 96 Stat. 2201 (1982) (codified at 42 U.S.C. §§ 10101–10226 [1982]. The 1982 statute has several provisions concerning low-level waste. See, e.g., 42 U.S.C. § 10171 (1982) (relating to financial requirements for the closure of low-level waste disposal sites).
2. P.L. 96–573, 94 Stat. 3347 (1980) (codified at 42 U.S.C. §§ 2021b-2021d [1982], as amended by P.L. 99–240, 99 Stat. 1842 (1986).
3. 42 U.S.C. § 2021d(a) (1982). See also "Low-Level Radioactive Waste Policy Act of 1985," U.S. House of Representatives, Report 99–314, Part 2, at 14 (October 1985).
4. U.S. Constitution, Article VI, Clause 2.
5. U.S. Constitution, Article I, Section 8, Clause 3. For another discussion of the effect of these two provisions on state authority to regulate radioactive

waste generated outside the state, see Campbell, W. A., "State Ownership of Hazardous Waste Disposal Sites: A Technique for Excluding Out-of-State Wastes?", *Envtl. L.* 14(1):177–196 (1983).

6. LLRWPAA § 3(a)(l), P.L. 99–240, 99 Stat.1843–44 (1986).
7. LLRWPAA § 4(a)(l), P.L. 99–240, 99 Stat. 1845 (1986).
8. LLRWPAA § 4(a)(2), P.L. 99–240, 99 Stat. 1845 (1986).
9. LLRWPAA § 4(c), P.L. 99–240, 99 Stat. 1846 (1986). See also "Low-Level Radioactive Waste Policy Act Amendments of 1985," U.S. Senate, Report 99-199, at 9 (November 1985).
10. LLRWPAA § 4(c), P.L. 99–240, 99 Stat. 1846 (1986).
11. LLRWPAA § 4(d), P.L. 99–240, 99 Stat. 1846 (1986).
12. U.S. Constitution, Article VI, Clause 2.
13. *Pacific Gas & Elec. Co.* v. *State Energy Resources Conservation and Dev. Comm'n*, 461 U.S. 190, 203–204 (1983).
14. Idem.
15. Idem at 204.
16. Idem.
17. Idem.
18. Idem at 212. For a general discussion of federal preemption of state regulation in the nuclear area, see Murphy, A. W., and La Pierre, D. B., "Nuclear 'Moratorium' Legislation in the States and the Supremacy Clause: A Case of Express Preemption," *Colum. L. R.* 76(3):392–456 (1976); Comment, "Federal Preemption of State Laws Controlling Nuclear Power," *Geo. L. J.* 64(6):1323–1361 (1976); Note, "Application of the Preemption Doctrine to State Laws Affecting Nuclear Power Plants," *Va. L. R.* 62(4):738–787 (1976).
19. 461 U.S. at 233. For a discussion of the remaining areas of state authority, see Comment, "Exercising Police Powers to Control Spent Fuel and Other Radioactive Wastes," *Golden Gate L. Rev.* 14(2):335–358 (1984).
20. *Illinois* v. *General Elec. Co.,* 683 F.2d 206 (7th Cir. 1982), cert. denied, 461 U.S. 913 (1983). See also *Jersey Central Power & Light Co.* v. *Township of Lacey*, 772 F.2d 1103 (3d Cir. 1985) (township's ban on the importation and storage of spent nuclear fuel and other radioactive waste generated outside the township is preempted by the Atomic Energy Act and the Hazardous Materials Transportation Act), cert. denied, 106 S. Ct. 1190 (1986).
21. *Illinois* v. *General Elec. Co.,* 683 F.2d at 215. See, generally, Kearney, R. C., and Garey, R. B., "American Federalism and the Management of Radioactive Wastes," *Pub. Ad. Rev.* 42(1):14–24 (1982).
22. *Washington State Bldg. and Constr. Trades Council* v. *Spellman*, 684 F.2d 627 (9th Cir. 1983), cert. denied, 461 U.S. 913 (1983), discussed in Case Note, *Gonz. L. Rev.* 18(3):605–626 (1982–83).
23. 684 F.2d at 630.
24. U.S. Constitution, Article I, Section 8, Clause 3.
25. See *Western & S. Life Ins. Co.* v. *Board of Equalization*, 451 U.S. 648, 652 (1981).
26. See *Philadelphia* v. *New Jersey*, 437 U.S. 617, 622 (1978).

27. Idem at 622–23; *General Elec. Co.*, 683 F.2d at 213.
28. 683 F.2d at 213.
29. See *Pike* v. *Bruce Church, Inc.*, 397 U.S. 137, 142 (1970).
30. *General Elec.*, 683 F.2d at 214; *Spellman*, 684 F.2d at 631.
31. *Spellman*, 684 F.2d at 631.
32. Idem. In 1984, 52% of the nation's low-level waste was disposed of at the site in Hanford, Washington. See "Low-Level Radioactive Waste Policy Act Amendments of 1985," U.S. House of Representatives, Report 99-314, Part 2, at 16 (October 1985).
33. See *Northeast Bancorp.* v. *Board of Governors*, 86 L. Ed. 2d 112, 125 (1985).
34. *Spellman*, 684 F.2d at 630.
35. See, e.g., Comment, "Congressional Recognition of State Authority Over Nuclear Power and Waste Disposal," *Chi.[-] Kent L. Rev.* 58(3): 813, 840, 844–45 (1982).
36. Idem at 845.
37. See, e.g., "Low-Level Radioactive Waste Policy Act Amendments of 1985," U.S. Senate, Report 99-199, at 10–11 (November 1985) (1985 amendments setting volume limitations on low-level waste to be disposed of at existing sites preempt conflicting state laws because Congress "does not intend to enact federal limits to disposal capacity in these states as part of the scheme set forth in this act only to find that the states themselves have established other limits under state law").
38. See Note, "Glowing Their Own Way: State Embargoes and Exclusive Waste Disposal Sites Under the Low-Level Radioactive Waste Policy Act of 1980," *Geo. Wash. L. Rev.* 53(3&4):654, 662 (1985).
39. Idem at 667 n. 164. For a discussion of Texas's efforts to establish its own low-level radioactive waste disposal site, see Colglazier, E.W., and English, M.R., "Low-Level Radioactive Waste: Can New Disposal Sites Be Found?" Tenn. L.R. 53(3): 621, 630–634 (1986).
40. See Note, "Glowing Their Own Way" (above, note 38), 676–79; Note, "Low-Level Radioactive Waste Disposal Compacts," *Va. J. Nat. Resources* 5(2): 382, 396–400 (1986). The Supreme Court recently narrowed the market participant exemption so that it does not cover state actions having "a substantial regulatory effect outside the particular market." *South Central Timber Dev., Inc.* v. *Wunnicke*, 467 U.S. 82, 97 (1984)
41. See *Env't Rep. (BNA) Curr. Dev.* 16(26):1088 (October 25, 1985).
42. See *New Jersey* v. *Nevada,* No. 104 -Orig. (U.S. Oct. 21, 1985); Env't Rep. (BNA) Curr. Dev. 16(26):1088 (October 25, 1985).
43. LLRWPAA § 4(c), P.L. 99-240, 99 Stat. 1846 (1986).
44. See, generally, Hardy, P. T., *Interstate Compacts: The Ties That Bind* (Athens, GA: Institute of Government, University of Georgia, 1982); Ridgeway, M. E., *Interstate Compacts: A Question of Federalism* (Carbondale, IL: Southern Illinois University Press, 1971); Thursby, V. V., *Interstate Cooperation: A Study of the Interstate Compact* (Washington, DC: Public Affairs Press, 1953); Frankfurter, F., and Landis, J. M., "A Study in

Interstate Adjustments," *Yale L. J.* 34(7):685–758 (1925); and Zimmermann, F. L., and Wendell, M., "The Interstate Compact and *Dyer* v. *Sims*," *Colum. L. R.* 51(8):931–950 (1951).

45. See, e.g., *Texas* v. *New Mexico*, 462 U.S. 554, 575 (1983); *Louisiana* v. *Texas*, 176 U.S. 1, 17 (1900).

46. *West Virginia ex rel. Dyer* v. *Sims*, 341 U.S. 22, 27 (1951).

47. See, e.g., *Green* v. *Biddle*, 21 U.S. (8 Wheat.) 1 (1823).

48. See, e.g., *Cuyler* v. *Adams*, 449 U.S. 433 (1981).

49. See, e.g., *Arizona* v. *California*, 373 U.S. 546 (1963).

50. See, e.g., *West Virginia ex rel. Dyer* v. *Sims*, 341 U.S. 22 (1951). For a list of the subjects covered by interstate compacts, see The Council of State Governments, "Interstate Compacts and Agencies" (1979); Hardy, *Interstate Compacts* (above, note 44), p. 5; and Thursby, *Interstate Cooperation* (above, note 44), pp. 97–124.

51. U.S. Constitution, Article I, Section 10, Clause 3.

52. See *Northeast Bancorp* v. *Board of Governors*, 86 L. Ed. 2d 112, 126 (1985); *Virginia* v. *Tennessee*, 148 U.S. 503, 518 (1893).

53. See *Florida* v. *Georgia*, 58 U.S. 478, 495 (1855).

54. See *Petty* v. *Tennessee-Missouri Bridge Comm'n*, 359 U.S. 275, 282 n. 7 (1959).

55. *Northeast Bancorp* v. *Board of Governors*, 86 L. Ed. 2d 112, 126 (1985); *Virginia* v. *Tennessee*, 148 U.S. 503, 519 (1893).

56. See *Northeast Bancorp*, 86 L. Ed. 2d at 126; *Virginia* v. *Tennessee*, 148 U.S. 503, 519 (1893).

57. See *United States Steel Corp.* v. *Multistate Tax Comm'n*, 434 U.S. 452, 468 (1978); *Virginia* v. *Tennessee*, 148 U.S. 503, 518 (1893).

58. See *Virginia* v. *Tennessee*, 148 U.S. 503, 519 (1893).

59. See, e.g., *Northeast Bancorp* v. *Board of Governors*, 86 L. Ed. 2d 112, 125–26 (1985); *United States Steel Corp.* v. *Multistate Tax Comm'n*, 434 U.S. 452 (1978).

60. *Washington State Bldg. and Constr. Trades Council* v. *Spellman*, 684 F.2d 627 (9th Cir. 1982), cert. denied, 461 U.S. 913 (1983).

61. See Peckinpaugh, T.L., "An Analysis of Regional Interstate Compacts for the Disposal of Low-Level Radioactive Wastes," *J. Energy L. & Pol.* 5(1):32 (1983) (citing Zimmermann, F. L., and Wendell, M., *The Law and Use of Interstate Compacts* [1961], 23).

62. See *Silkwood* v. *Kerr-McGee Corp.*, 104 S. Ct. 615, 622 (1984); *Pacific Gas & Elec. Co.* v. *State Energy Resources Conservation & Dev. Comm'n*, 461 U.S. 190, 205 (1983).

63. "Low-Level Radioactive Waste Policy Act Amendments of 1985," U.S. House of Representatives, Report 99–199, at 9 (November 1985).

64. See *Green* v. *Biddle*, 21 U.S. (8 Wheat.) 1, 85–86 (1823).

65. See *Cuyler* v. *Adams*, 449 U.S. 433, 441 (1981); *Virginia* v. *West Virginia*, 78 U.S. (11 Wall.) 39, 59 (1871).

66. See *Cuyler* v. *Adams*, 449 U.S. 433, 441 (1981); *Virginia* v. *Tennessee*, 148 U.S. 503, 521 (1893).

67. LLRWPAA § 4(c)(2), P.L. 99-240, 99 Stat. 1846 (1986).
68. *James* v. *Dravo Contracting Co.*, 302 U.S. 134, 148 (1937).
69. LLRWPAA § 4(d), P.L. 99-240, 99 Stat. 1846 (1986).
70. "Low-Level Radioactive Waste Policy Act Amendments of 1985," U.S. House of Representatives, Report 99-314, Part 2, at 29 (October 1985).
71. See *United States* v. *Tobin*, 306 F.2d 270, 272-74 (D.C. Cir.), cert. denied, 371 U.S. 902 (1962). See also *Riverside Irrigation Dist.* v. *Andrews*, 568 F. Supp. 583, 589 (D. Colo. 1983), aff'd, 758 F.2d 508 (10th Cir. 1985).
72. See, e.g., Peckinpaugh, "An Analysis of Regional Interstate Compacts" (above, note 61), 33 and 36-37, and King, D. B., "Interstate Water Compacts," in *Water Resources and the Law*, edited by The University of Michigan Law School (Ann Arbor, MI, 1958), p. 362.
73. See LLRWPAA §5(a) and (b), P.L. 99-240, 99 Stat. 1846-49 (1986).
74. "Low-Level Radioactive Waste Policy Act Amendments of 1985," U.S. House of Representatives, Report 99-314, Part 1, at 21 (October 1985).
75. See *Petty* v. *Tennessee-Missouri Bridge Comm'n*, 359 U.S. 275, 281-82 (1959). See also Peckinpaugh, "An Analysis of Regional Interstate Compacts" (above, note 61), 33-35, and Zimmerman and Wendell, "The Interstate Compact" (above, note 44), at 943-44 n. 66.
76. *Carchman* v. *Nash*, 105 S. Ct. 3401, 3403 (1985); *Lessee of Marlatt* v. *Silk*, 36 U.S. (11 Pet.) 1, 22-23 (1837).
77. See *Petty* v. *Tennessee-Missouri Bridge Comm'n*, 359 U.S. 275, 278 n. 4 (1959); see also *Delaware River Joint Toll Bridge Comm'n* v. *Colburn*, 310 U.S. 419, 431 (1940).
78. P.L. 99-240, §225, Art. VIIb, 99 Stat. 1900 (1986).
79. U.S. Constitution, Article I, Section 10, Clause 1. See King, "Interstate Water Compacts" (above, note 72), p. 365.
80. See Peckinpaugh, "An Analysis of Regional Interstate Compacts" (above, note 61), 38-39. See also King, "Interstate Water Compacts" (above, note 72), pp. 366-67.
81. Compare LLRWPAA, § 2(9), P.L. 99-240, 99 Stat. 1842 (1986) with P.L. 99-240, § 221, Art. II(2), 99 Stat. 1861 (1986). See also Peckinpaugh, "An Analysis of Regional Interstate Compacts" (above, note 61), 39-40.
82. LLRWPAA § 223, Art. VI(A)(1)-(2), P.L. 99-240, 99 Stat. 1877-78 (1986).
83. LLRWPAA § 4(b)(4), P.L. 99-240, 99 Stat. 1846 (1986). See also LLRWPAA § 4(b)(3), P.L. 99-240, 99 Stat. 1845-46 (1986).
84. *Green* v. *Biddle*, 21 U.S. (8 Wheat.) 1, 92 (1823).
85. *Virginia* v. *Tennessee*, 148 U.S. 503, 520 (1893).
86. LLRWPAA § 221, Art. IV(5), P.L. 99-240, 99 Stat. 1862 (1986).
87. See *Philadelphia* v. *New Jersey*, 437 U.S. 617 (1978).
88. See, e.g., LLRWPAA § 223, Art. 5A, P.L. 99-240, 99 Stat. 1877 (1986).
89. *West Virginia ex rel. Dyer* v. *Sims*, 341 U.S. 22, 28 (1951).
90. *Texas* v. *New Mexico*, 462 U.S. 554, 564 (1983).
91. See Thursby, *Interstate Cooperation* (above, note 44), p. 79, and King, "Interstate Water Compacts" (above, note 72), p. 374.
92. See Thursby, *Interstate Cooperation* (above, note 44), p. 93.

93. See Thursby (idem), p. 89, and King, "Interstate Water Compacts" (above, note 72), pp. 374–75.

94. See "Low-Level Radioactive Waste Policy Act Amendments of 1985," U.S. House of Representatives, Report 99–314, Part 2, at 17–18 (October 1985).

95. See *Env't Rep. (BNA) Curr. Dev.* 16(26):1088 (Oct. 25, 1985).

96. See Thursby, *Interstate Cooperation* (above, note 44), p. 79.

97. Idem.

98. See, e.g., *Cooper* v. *Aaron*, 358 U.S. 1 (1958). See also the same case at 21–22 (Justice Frankfurter concurring—although the "use of force to further obedience to law is . . . a last resort, and one not congenial to the spirit of our Nation," the states must yield to the paramount authority of the federal government).

99. Idem at 18 (quoting *United States* v. *Peters*, 9 U.S. [5 Cranch] 115, 136 [1809]). See also *United States* v. *Louisiana*, 364 U.S. 500 (1960); *Rollins Envtl. Serv., Inc.* v. *Parish of St. James*, 775 F.2d 627, 637–38 (5th Cir. 1985).

100. See King, "Interstate Water Compacts" (above, note 72), pp. 375 and 410.

101. "Low-Level Radioactive Waste Policy Act Amendments of 1985," U.S. House of Representatives, Report 99–314, Part 1, at 21 n. 1 (October 1985).

102. See *Wharton* v. *Wise*, 153 U.S. 155, 171 (1894). But cf. *Virginia* v. *West Virginia*, 78 U.S. (11 Wall.) 39, 55 (1871).

103. LLRWPAA § 4(c)(2), P.L. 99–240, 99 Stat. 1846 (1986).

104. See "Low-Level Radioactive Waste Policy Act Amendments of 1985," U.S. House of Representatives, Report 99–314, Part 2, at 29 (October 1985).

105. See LLRWPAA §§ 221–227, P.L. 99–240, 99 Stat. 1860–1924 (1986). For a more recent survey of the compact configurations, see Colglazier and English, "Low-Level Radioactive Waste" (above, note 39), 623n.2.

106. See, e.g., LLRWPAA § 221, Art. VI(l), P.L. 99–240, 99 Stat. 1863 (1986), and § 222, Art. VIIe, P.L. 99–240, 99 Stat. 1870 (1986).

107. LLRWPAA § 223, Art. VIID.3, P.L. 99–240, 99 Stat. 1879 (1986).

108. See LLRWPAA, §§ 221–227, P.L. 99–240, 99 Stat. 1860–1924 (1986).

109. See, e.g., LLRWPAA § 221, Art. VI, P.L. 99–240, 99 Stat. 1863 (1986); § 222, Art. VII, P.L. 99–240, 99 Stat. 1869 (1986); § 223, Art. VII, P.L. 99–240, 99 Stat. 1878–80 (1986); and § 225, Art. VIII, P.L. 99–240, 99 Stat. 1900–01 (1986).

110. See *Texas* v. *New Mexico*, 462 U.S. 554, 564 (1983); and above, text accompanying note 90.

111. See above, notes 70–71 and accompanying text.

112. See "Low-Level Radioactive Waste Policy Amendments of 1985," U.S. Senate, Report 99–199, at 5 (November 1985); Colglazier and English, "Low-Level Radioactive Waste" (above, note 39) 623n.2 and 630–637. Note, "Glowing Their Own Way" (above, note 38), 656 n. 16.

113. See Note, "Glowing Their Own Way" (above, note 38), 656, and 668 n. 98.

114. See idem, 656 n. 14.

115. See idem, 665 n. 79.

116. See LLRWPAA § 5(e), P.L. 99–240, 99 Stat. 1852–54 (1986).

117. See Note, "Glowing Their Own Way" (above, note 38), 668 n. 98.

118. LLRWPAA § 3(a)(l), P.L. 99–240, 99 Stat. 1843 (1986) (emphasis added).

119. See Note, "Glowing Their Own Way" (above, note 38), 667.

120. LLRWPAA § 4(c), P.L. 99–240, 99 Stat. 1846 (1986) (emphasis added).

121. LLRWPAA § 2(4), P.L. 99–240, 99 Stat. 1842 (1986).

122. See Note, "Glowing Their Own Way" (above, note 38), 664, and 665 n. 79.

123. "Nuclear Waste Policy Act," U.S. House of Representatives, Report 96-1382, Part 1, at 34–35 (September 1980). See Note, "Glowing Their Own Way" (above, note 38), 667.

124. "Low-Level Radioactive Waste Policy Act Amendments of 1985," U.S. House of Representatives, Report 99-314, Part 2, at 22 (October 1985).

125. Cf. *New England Power Co.* v. *New Hampshire*, 455 U.S. 331 (1982). But see above, text accompanying notes 39–40, for a discussion of the market participant theory of authorized interstate discrimination.

126. See, e.g., *Rollins Envtl. Serv., Inc.* v. *Parish of St. James*, 775 F.2d 627 (5th Cir. 1985).

CHAPTER 4

Low-Level Radioactive Waste Disposal

Frank L. Parker

Before one can fully comprehend all of the social and political problems created by the disposal of low-level radioactive waste, there must be a factual basis upon which to build such an understanding. In this chapter, some answers are presented to questions about low-level wastes, what they are, who produces them, and past, present, and likely future means of dealing with such wastes.

WHAT ARE LOW-LEVEL WASTES?

There are many definitions of radioactive wastes, ranging from that of the International Atomic Energy Agency (IAEA)[1] to those of individual laboratories. In the U.S.A., low-level radioactive wastes are defined by law. Low-level waste is defined as "radioactive waste not classified as high-level radioactive waste, transuranic waste, spent nuclear fuel, or by-product materials as defined in Section 11e(2) of the Atomic Energy Act (uranium or thorium tailings and wastes)."[2]

In 10 CFR 61, "Licensing Requirements for Land Disposal of Radioactive Waste" (low-level waste only), the definition states that "waste means those low-level radioactive wastes containing source, special nuclear, or by-product materials that are acceptable for disposal in a land disposal facility. For purpose of this definition, low-level waste has the same meaning as in the Low-Level Radioactive Waste Policy Act [LLRWPA]."[3]

The law, therefore, tells us what low-level wastes are not. According to the Nuclear Waste Policy Act of 1982,[4] "the term 'high-level radioactive waste' means (A) the highly radioactive material resulting from the reprocessing of spent nuclear fuel, including liquid waste produced

Low-Level Radioactive Waste Regulation: Science, Politics, and Fear, Michael E. Burns, Ed., © 1988 Lewis Publishers, Inc., Chelsea, Michigan—Printed in USA.

directly in reprocessing and any solid material derived from such liquid waste that contains fission products in sufficient concentrations; and (B) other highly radioactive material that the Commission, consistent with existing law, determines by rule requires permanent isolation" (Section 2[12]), and "the term 'low-level radioactive waste' means radioactive material that (A) is not high-level radioactive waste, spent nuclear fuel, transuranic waste or by-product material as defined in Section 11e(2) of the Atomic Energy Act of 1954;[5] and (B) the Commission, consistent with existing law, classifies as low-level radioactive waste." Transuranic waste is defined in 40 CFR 191:[6] " 'transuranic radioactive waste' as used in this part, means waste containing more than 100 nanocuries of alpha-emitting transuranic isotopes, with half-lives greater than twenty years, per gram of waste, except for: (1) high-level radioactive waste; (2) wastes that the Department has determined, with the concurrence of the Administrator, do not need the degree of isolation required by this part; or (3) wastes that the Commission has approved for disposal on a case-by-case basis in accordance with 10 CFR 61."

10 CFR 61 "contains specific requirements for near-surface disposal of radioactive waste which involves disposal in the uppermost portion of the earth, approximately 30 meters," and that "site characteristics should be considered in terms of the indefinite future and evaluated for at least a 500 year time frame."[7] In 10 CFR 61, low-level wastes are divided into three classes, A, B, and C, based upon their long- and short-lived constituents. Numerical limits for Class A, B, and C low-level wastes are determined by the concentrations of long-lived radionuclides, as shown in Table 1,[8] and concentrations of short-lived radionuclides, shown in Table 2.[9] Class A wastes are defined as "waste that does not contain sufficient amounts of radionuclides to be of concern" with respect to migration of radionuclides, long-term active maintenance, and potential exposures to intruders, and that "tends to be stable, such as ordinary trash-type wastes."[10] If the concentration does not exceed 0.1 of the value in the table, it is Class A waste; if the concentration exceeds 0.1 of the value, but is less than the value in the table, it is Class C waste; and if it exceeds the concentration in Table 1, the waste "is not generally acceptable" for near-surface disposal. For mixtures of such wastes, the standard sum of fractions of the concentration of each of the contained wastes to the concentrations in Table 1 shall not exceed the limits indicated above.

If the wastes do not contain any of the radionuclides listed in Table 1, then they are classified by their short-half-lived components, shown in Table 2. If the concentrations do not exceed those listed in Column 1, the waste is Class A. If the concentrations are greater than those in Column 1, but equal to or less than those in Column 2, the waste is Class B. If the concentration is greater than the values in Column 2 and equal to or less

Table 1. Long-Lived Material

Radionuclide	Concentration
C-14	8 Ci/m^3
C-14 in activated metal	80 Ci/m^3
Ni-59 in activated metal	220 Ci/m^3
Nb-94 in activated metal	0.2 Ci/m^3
Tc-99	3 Ci/m^3
I-129	0.08 Ci/m^3
Alpha-emitting transuranic nuclides with half-life >5 yr	100 nCi/g
Pu-241	3,500 nCi/g
Cm-242	20,000 nCi/g

Table 2. Short-Lived Material

| Radionuclide | Concentration, Ci/m^3 | | |
	Column 1	Column 2	Column 3
Total of all nuclides with half-life <5 yr[a]	700	—	—
H-3[a]	40	—	—
Co-60[a]	700	—	—
Ni-63	3.5	70	700
Ni-63 in activated metal	35	700	7000
Sr-90	0.04	150	7000
Cs-137	1	44	4600

[a]There are no limits established for these radionuclides in Class B or C wastes. Practical considerations such as the effects of external radiation and internal heat generation on transportation, handling, and disposal will limit the concentrations for these wastes. These wastes shall be Class B unless the concentrations of other nuclides in the table determine the waste to be Class C independent of these nuclides.

than those in Column 3, the waste is Class C. If the concentrations are greater than the values in Column 3, "the waste is not generally acceptable for near-surface disposal." For mixtures of Table 2 wastes, the sum of fractions rule holds. For wastes containing both long- and short-lived wastes, wastes classified as Class A wastes by Table 1, the classifications shall be determined by wastes listed in Table 2. If wastes are classified as Class C by Table 1, then they shall be classified as Class C, provided that the concentration of wastes listed in Table 2 does not exceed the values in Column 3.

If the radioactive waste contains only radionuclides not listed in either

Table 1 or 2, it is classified as Class A. Because of their greater hazard potential, Class B wastes must meet more rigorous requirements for waste form stability than do Class A wastes. Class C wastes must also meet more rigorous requirements for waste form, and have further restrictions on their burial, such as a minimum depth of 5 m below the top surface of the cover, or with barriers designed to protect against inadvertent intrusion for at least 500 years. Wastes that do not meet the requirements of Class C for near-surface disposal require waste forms and disposal methods that are "different and in general more stringent than those specified for Class C waste."[11]

Most other countries are now using the special names for some units of the Système International d'Unités (SI) used in the field of ionizing radiation. The gray (Gy) is used for absorbed dose and is equal to 1 joule/kg, or 100 rads. The becquerel (Bq) is used for the activity of a radionuclide and is equal to the quantity of the radionuclide undergoing one disintegration in 1 sec, or 2.7×10^{-11} curies (Ci). The sievert (Sv) is the unit for radiation dose equivalent used for radiation protection purposes and is equal to the gray times the quality factor times the distribution factor, or 100 rem.

Though this country has not yet officially adopted this system, the National Council on Radiation Protection and Measurements (NCRP) has adopted it, and the federal government is moving toward such adoption. Consequently, the units used in this chapter will be those used in the original. The conversions are shown in the appendix to this chapter.

Waste classification, both in the U.S.A. and in other countries, has become important because disposal limitations are based upon the classification. There have been recent proposals in the U.S.A. to classify all wastes with concentrations greater than Class C as high-level waste and, thereby, require deep geologic disposal. In the United Kingdom, the decision has been made recently to require that any wastes greater in concentration than very low-level waste (VLLW) be disposed of in other-than-unengineered near-surface disposal (VLLW is defined as < 400 kBq in beta/gamma activity and < 0.1 m³ in volume, or < 40 kBq beta/gamma for single items).

One of the reasons for the interest in classification is that a group of wastes has escaped classification. First, there are "de minimis" wastes that are of such low concentration and pose such low risks (negligible annual dose equivalents) that they are below regulatory concern (BRC). Numerous proposals have been made to establish such levels at dosages ranging from 0.001 rem/yr to concentrations below 1 μCi/cm³.[12]

The Nuclear Regulatory Commission (NRC) has, in fact, already ruled that animal carcasses and scintillation fluids are below regulatory concern and can be disposed of without regard to radioactivity if the concentration is ≤ 0.05 μCi/g of H-3 or C-14 (10 CFR 20.306). Radioactive wastes can be disposed of in sewers, if (1) amounts ≤ 1 Ci/yr of C-14 and ≤ 5 Ci of H-3/yr are released to sewers by the licensee; (2) amounts released in any one day are ≤ 10 times the quantity of radionuclides shown in Appendix C of 10 CFR 20, or the daily amounts, based on full dilution on a daily basis with the sewage released into the sewer by the licensee, are within the concentration limits stated in Appendix B of 10 CFR 20; (3) the quantity released in any one month, if diluted by average monthly quantity released by the licensee, will not exceed the limits specified in Appendix B of 10 CFR 20; and (4) licensee remains within limits of 5 Ci/yr of H-3, 1 Ci/yr of C-14, and 1 Ci/yr for all other isotopes.

It has been estimated that such releases to water or soil would result in a maximum individual dose of 1 mrem/yr or, if biomedical wastes are incinerated that contain 6 Ci of C-14, they will result in 0.6 health effects/1,000 generations or 0.067 health effects per Ci released.[13]

The International Commission on Radiological Protection[14] recently recommended exemption rules, i.e., a dose below which there is "no further need for radiation protection concern," because the cost or impacts of reducing the dose below those levels would be greater than the impact from the dose itself. This exemption level is set at an annual individual dose equivalent of < 0.01 mSv and a collective dose commitment lower than a man-Sv caused by a practice or source over a defined period of operation.

In the Final Environmental Impact Statement (FEIS) for 10 CFR 61,[15] 37 waste streams were analyzed for their volumes and physical, chemical, and radiological properties that were projected to be routinely generated from 1980 to 2000. It was estimated that perhaps of these waste streams, Fuel Fabrication Noncompactible Trash (F-NCTRASH), Fuel Fabrication Compactible Trash (F-COTRASH). Fuel Fabrication Process Wastes (F-PROCESS), Industrial Source and Special Nuclear Material Waste (N-SSWASTE), Industrial Source and Special Nuclear Material Trash (N-SSTRASH) (large facilities), and UF$_6$ Process Waste (U-PROCESS) could be classified as de minimis (doses < 1 mrem/yr). According to the preliminary calculations of the Environmental Protection Agency (EPA), if this were done, the savings could range from $60 to $110 million per year.[12] It is estimated that $13 million per year has already been saved by allowing conventional disposal of biomedical wastes rather than disposal in a licensed low-level waste facility.[13]

The change from 10 to 100 nCi/g for the definition of Transuranic Waste (TRU) has resulted in an estimated total savings of $250,000,000 (in 1982 dollars — $460/m^3 for shallow land burial and $20,000/m^3 for disposal in a geological repository).[16] Further, an analysis of the effects of an increase in the limit for the classification of transuranic wastes

Table 3.   Incremental Impact for Transuranic Waste Disposal at
100 nCi/g

	Incremental Dose mrem/yr per person	
	Pu-238	Pu-239
Groundwater as drinking water	10^{-3}	10^{-4}
Site boundary analysis		
Savannah River as drinking water	10^{-5}	10^{-6}
Human food chain scenario	10^{-4}	10^{-5}
Recreational uses of river water	10^{-8}	10^{-9}
Hypothetical land occupation scenario		
Base case: 100 years in future	10^{-1}	10^{-2}
Lower leach rate based on		
lysimeter data	10^{-2}	10^{-3}
No vegetative uptake (Greater		
Confined Disposal or overburden)	10^{-3}	10^{-4}

from 10 to 100 nCi/g for a variety of scenarios of disposal at the Savannah River Plant, Aiken, South Carolina, resulted in a maximum increase in dose of 0.1 mrem/yr and ranged down to 10^{-9} mrem/yr, as shown in Table 3.[17]

The president has ordered[18] that high-level defense waste and high-level commercial waste should be disposed of in the same geologic repository. The wastes that have concentrations greater than Class C (10 CFR 61) and are not included within the waste streams defined as high-level waste in the Nuclear Waste Policy Act are left in limbo and must be treated case by case. Not only do a number of defense waste streams fall within this category, but so do special commercial waste streams, such as the ion exchange resins from the cleanup at Three Mile Island. Natural and accelerator-produced materials are not regulated under the NRC and should, therefore, be regulated by the EPA.

Finally, there are the ubiquitous mixed wastes—wastes containing both radioactive materials and hazardous chemical waste materials.[19,20] Commercial wastes within this category are wastes without a home. No commercial hazardous waste disposer will handle them, as there are no applicable regulations, nor is it certain who will have ultimate authority to handle them—NRC (for nuclear materials), under 10 CFR 61, or EPA (for hazardous materials), under 40 CFR 264.

To complicate matters even further, the Department of Energy (DOE) has the authority to dispose of its own low-level radioactive waste without licensing by the NRC. High-level waste disposal, both commercial and defense, must be licensed by the NRC. Finally, to cap this "mystery wrapped in an enigma" tale, EPA is planning to issue regulations on low-level radioactive wastes that almost certainly will not be fully compatible with the regulations of the NRC. Under regulations of both NRC and EPA, there can be "agreement states," where licensing and control of low-level waste and hazardous waste disposal can be delegated to the states — which has already occurred in many states. The consequences of such allocations are not just turf battles for power and money, but highlight the radically different requirements for disposal of low-level radioactive wastes and hazardous wastes. As has previously been stated, Class C low-level wastes must provide for 500 years of intruder protection and 100 years of institutional control, while hazardous wastes do not require intruder protection, and after 30 years, the sites can be abandoned, unless the EPA's Regional Administrator rules otherwise. In addition, there are very specific regulations concerning linings for the disposal sites and for groundwater monitoring in the Resource Conservation and Recovery Act (RCRA), but not in 10 CFR 61.[21] 10 CFR 61 allows controlled releases into the environment; RCRA does not. The 1984 amendments to RCRA state that near-surface ground disposal "should be the least favored method for managing hazardous waste."[22]

As can be seen, none of these definitions has anything to do explicitly with the risk to humans or the environment. If our aim is to allocate risks equally (which may *not* be our aim), then rules and regulations that do not take into account whether the wastes are disposed of in the humid or arid parts of the country, or in populated or barren areas, do not make much sense. In addition, as William D. Ruckelshaus has pointed out, sound public policy would "first, balance risks against the local economic impacts of controlling them; second, ensure that our national programs that attempt to deal with local risks operate according to risk management principles; and third, involve the local public in a meaningful way in the decision-making process."[23] Despite this advice from the man who has twice headed EPA, none of these recommendations are being implemented.

However, an attempt is being made to establish waste categories based on risk. NCRP wrote a letter to DOE pointing out that waste categories should be risk based.[24] (It should be noted that basing waste categories on risk locks one into the present technology.)

Finally, there have been only six operating commercial low-level waste burial grounds in this country, and only three are currently in operation. When the burial grounds mandated by the Low-Level Radioactive Waste Policy Act are operational, there are likely to be fewer than 10 in operation at any time. It seems, therefore, that it is inappropriate to use a generic site for generic wastes to derive methods of treatment, packaging, and disposal and then allow exceptions, as in 10 CFR 61. Since sites are so few, one would be far better off to evaluate each site for its capabilities and then set treatment, packaging, and disposal requirements suitable for that specific site. A risk-based performance standard for each site, not a quantity- or concentration-based standard for a generic site applied to a specific site, is the most appropriate approach.

WHERE DO THE WASTES COME FROM?

Perhaps the best description of low-level waste is given in the FEIS for 10 CFR 61. Low-level radioactive wastes can be solid, liquid, or gaseous and can be generated by industries such as "hospitals; medical, educational or research institutions; private or government laboratories; or facilities forming part of the nuclear fuel cycle (e.g., nuclear power plants, fuel fabrication plants)."[15] The NRC has estimated that about 85,000 m³ of low-level waste are generated and disposed of at commercial low-level waste disposal sites annually. It is projected that about 3.6 × 10⁶ m³ will be generated between 1980 and 2000. Though over 20,000 groups are licensed to use radioactive materials, most of the radioactivity disposed of at commercial disposal sites is generated by fewer than 100 licensees.

The low-level wastes considered in NRC's analysis can be divided into four groups: (1) light water reactor process wastes, including ion exchange resins, concentrated liquids, filter sludges, and filter cartridges; (2) trash from reactors, fuel fabrication, and institutional and industrial facilities; (3) low specific activity wastes, from fuel fabrication, UF_6 processes, and institutional and industrial sources; and (4) special wastes, including reactor nonfuel core components, decontamination resins, waste from isotope production facilities, accelerator targets, sealed sources, industrial high-activity waste, and mixed oxide facility decontamination wastes.

These wastes are a very heterogeneous group, with their radioactive content ranging from negligible to highly hazardous. A better picture of what goes into a low-level waste disposal facility can be gained from looking at the actual receipt stream at U.S. Ecology's site in Richland, Washington,[25] just across the road from the defense waste storage and burial sites. The types of wastes are shown in Table 4; no distinction in analysis was made between combustible and noncombustible wastes, and containers with external radiation levels of > 200 mrad/hr and with high concentrations of tritium (> 13.3 mCi/ft³) were not inventoried. It is estimated that 32% to 44% of the waste is combustible. It is clear that trash was the major contributor by volume.

Table 4.   Characteristics of Waste Shipped to Richland Site, 1983

Waste Type	Percent by Volume
Vials	4.2
Solid dry (dry active wastes)	82.8
Solidified liquids	4.4
Biological	1.6
Absorbed liquids	5.4
Resin materials	1.2
Filter media	0.4

The external radiation levels are shown in Table 5. It is clear that the majority of containers have very low surface doses (54% are < 2 mrad/ hr — which means an occupational worker could sit on them all day every working day and receive a dose < 100 mrem/wk — a permissible, though not a desirable, level) with a small but significant fraction (8%) having surface readings > 100 mrad/hr. Almost half (46%) of the containers received were the standard 55-gal drums, and almost all the rest of the waste was in 50- to 100-ft³ boxes or liners (31%) or 100- to 200-ft³ boxes or liners (17%). Most (56%) were slightly compacted, to < 33.3 lb/ft³ (the density of truck-compacted municipal refuse is about 20 lb/ft³), while only 28% was compacted to significantly higher density (> 46.8 lb/ft³). The concentration of isotopes and the frequency with which they occur in the dry active waste (DAW) is shown in Table 6 and compared to typical power plant trash constituents. Though Co-60 dominates (78%) the radioactive content of power plant waste, tritium and C-14 contribute 75% of the radioactive content of the waste at U.S. Ecology. Annual disposal costs (in 1983 dollars) for low-level waste from an average nuclear power plant are $750,000 for a pressurized water reactor (PWR) and $2,000,000 for a boiling water reactor (BWR).

For another perspective, wastes expected to be received at the Oak Ridge Reservation's Proposed Central Waste Disposal Facility (now withdrawn) are examined.[26] The wastes are low-level wastes from the three plants located on the Oak Ridge Reservation: (1) the Y-12 Production Plant, which produces nuclear weapons components, process, source, and special nuclear material, and provides support to weapons-design laboratories and other government agencies; (2) the Oak Ridge Gaseous Diffusion Plant (ORGDP), now on standby, which previously had as its primary purpose the enrichment of uranium hexafluoride in U-235 (in addition, it had extensive R & D efforts underway on all three methods of enrichment [laser isotopic, gas centrifuge, and gaseous diffusion]; only laser isotopic enrichment research continues); and (3) Oak

Table 5. Surface Radiation Readings of Containers Shipped to
Richland Site, 1983

Radiation Reading (mR/hr)	Percent by Volume
0.0 to 0.20	21.8
0.21 to 1.00	23.7
1.01 to 2.00	8.4
2.01 to 5.00	9.7
5.01 to 10.00	6.8
10.01 to 20.00	6.5
20.01 to 40.00	6.6
40.01 to 60.00	4.3
60.01 to 80.00	2.5
80.01 to 100.00	1.8
Over 100.00	7.9

Table 6. Dry Active Waste

Isotope	Activity Typical Power Plant (mCi/lb)	Activity U.S. Ecology Analysis (mCi/lb)	Isotope Frequency (%)
Co-60	.03400	.00580	65.0
H-3	.00085	.02430	17.4
C-14	.00003	.01401	12.0
I-131	---	.00310	14.4
Cs-137	.00878	.00396	63.0

Ridge National Laboratory (ORNL), a large multipurpose research laboratory. Total employment at the three plants was about 15,000.

The three plants were expected to generate routinely about 11,000 m³/
yr of solid low-level radioactive waste. The low-level waste is about 90%
of the total volume generated at the three, but has only a few percent of
the activity and would, therefore, generally, be classified by NRC as
Class A. The types of waste and the composition to be expected from
each plant are shown in Table 7.

Table 7. Characteristics of Solid Debris Low-Level Waste

Waste Parameter	ORNL[a]		Y-12[a]	ORGDP[a]	
	Baseline	Nonroutine	Baseline	Baseline	Nonroutine
Volume Rate[b]	1.7×10^3 m³/yr (5.9×10^4 ft³/yr)	1.1×10^3 m³/yr (4.0×10^4 ft³/yr)	1.8×10^3 m³/yr (6.3×10^4 ft³/yr)	1.0×10^3 m³/yr (3.5×10^4 ft³/yr)	7.6×10^3 m³/yr (2.7×10^5 ft³/yr)
% Combustible	30	—	~30	—	—
Average Isotopic Conc., Ci/m³					
H-3	1×10^{-1}	—	—	—	—
C-14	1×10^{-3}	—	—	—	—
Co-60	1×10^{-3}	—	—	—	—
Sr-90	2×10^{-2}	—	—	—	—
Zr-93	1×10^{-3}	—	—	—	—
Tc-99	—	—	—	3×10^{-3}	2×10^{-2}
Sn-121	6×10^{-4}	—	—	—	—
Cs-134	1×10^{-3}	—	—	—	—
Cs-137	5×10^{-3}	—	—	—	—
Sm-151	1×10^{-3}	—	—	—	—
Ir-192	2×10^{-3}	—	—	—	—
U-234	—	—	6×10^{-5}	7×10^{-5}	5×10^{-5}
U-235	—	—	2×10^{-3}	4×10^{-6}	2×10^{-6}
U-238	—	—	—	5×10^{-5}	4×10^{-5}
Pu-238	—	3×10^{-5}	—	—	—
Pu-239	—	9×10^{-6}	—	—	—

Continued on next page

Table 7. *Continued*

Waste Parameter	ORNL[a] Baseline	ORNL[a] Nonroutine	Y-12[a] Baseline	ORGDP[a] Baseline	ORGDP[a] Nonroutine
Am–241	—	4×10^{-5}	—	—	—
Cm–244	—	4×10^{-5}	—	—	—
Mixed TRU (mostly Pu–239)	—	9×10^{-5}	—	—	—
Components	Paper, cloth, plastics, rubber, wood, metals, glass, ceramics, concrete, soil resins	Concrete and other building materials (D&D wastes)	Baled and bulk	Miscellaneous materials plus some incinerator ash	Scrap metal[c]

[a]Baseline wastes are received on a regular schedule; nonroutine wastes are received on a campaign schedule.
[b]The rate of receipt of solid debris LLW from ORNL is expected to begin at about 2×10^3 m³/yr and may increase, over a five-year period, to about 3×10^3 m³/yr (9×10^4 ft³/yr).
[c]ORGDP will produce 10,000 tons of scrap metal during 1985 only.

WHERE DO THE WASTES GO?

As was pointed out earlier, the wastes go and have gone to sites shown in Table 8. The difficulties or lack of difficulties at each site will be explored in this section. The U.S. Geological Survey (USGS) has carried out comprehensive studies of the geology, hydrology, and radionuclide transport at all of the commercial sites except U.S. Ecology's Hanford site. They have also studied the Idaho National Engineering Laboratory, Oak Ridge National Laboratory, Argonne National Laboratory, Hanford Reservation, and the Nevada Test Site. An evaluation and summary of the results have been published.[27,28] The three commercial sites closed were in the humid region of the country. The technical results and problems at all three sites were similar, in that they were hydrogeologically complex, suffered from surface cap collapse, and allowed water to infiltrate and collect in the trenches.

Surprisingly, the one remaining operating humid site, Barnwell, has not had these difficulties. The USGS will continue to study the site to learn the reasons for its successful performance. The USGS has also turned its attention to more basic topics of scientific interest, such as unsaturated hydrology, geochemistry, clay mineralogy, surface geophysical techniques, and model development and testing.

Design problems have been tabulated by Oztunali[29] and are shown in Table 9. Some of the layout problems were: *geometry*—disposal cells frequently dug too close to each other, or with the bottom elevation at one end above the top elevation at the other end; *man-made impositions*—insufficiently compacted soil at one end and wastes placed in it; *geologic features*—trenches parallel to strike with trench wall stability problems.

Problems with disposal cell preparation were: *drainage systems*—poor surface water drainage and failure of subsurface drainage systems; *heterogeneities*—unpredictable performance because of stratigraphic and lithological heterogeneities; *man-made*—trench filled in to above actual ground water rather than above highest to be expected; *soil properties*—trench walls too steep, with resulting stability problems.

Problems with waste emplacement and backfill were: *standing meteoric water*—disposal of wastes into such water or closing trenches under such conditions; *backfill materials properties*—usually excavated material with no consideration for its ability to fill void spaces; *backfill practices*—void spaces and trench collapse with segments of trench being isolated; *sumps*—not performing as expected because of inadequate drainage and design.

Problems with disposal cell cover and stabilization were: *cover/bottom/side permeability*—cover more permeable, with "bathtub" effect

Table 8. Commercial Waste Disposal Sites

Location	Operator	Originally Licensed by (Year)	Currently Licensed by	Operational Status
Beatty, Nevada	U. S. Ecology, Inc.	AEC (1962)	State	Open
Maxey Flats, Kentucky	U. S. Ecology, Inc.[a]	Kentucky (1962)	State	Closed
West Valley, New York	Nuclear Fuel Services, Inc.	New York (1973)	State	Closed
Richland, Washington	U. S. Ecology, Inc.	AEC (1965)	State and NRC[b]	Open
Sheffield, Illinois	U. S. Ecology, Inc.	AEC (1967)	NRC	Closed
Barnwell, South Carolina	Chem-Nuclear Systems, Inc.	South Carolina (1971)	State and NRC[b]	Open

[a]U. S. Ecology was the operator while the site was open. Currently, Hittman, Inc. maintains the site as a caretaker for the state of Kentucky.
[b]NRC licenses only special nuclear material.

Table 9.   Summary of Design-Related Negative Past Experiences

Disposal Site Layout

Geometry of the disposal cells
Man-made impositions
Consideration of geologic features

Disposal Cell Preparation

Drainage systems
Heterogeneities
Man-made impositions
Soils properties

Waste Emplacement and Backfill

Standing meteoric water
Backfill material properties
Backfill practices
Sumps

Disposal Cell Cover and Stabilization

Cover vs bottom/side permeability
Vegetation
Man-made impositions
Waste stability

---

developing; *vegetation* — cells frequently poorly vegetated and with little thought to biotic transport by vegetation; *man-made impositions* — fill placed over entire area in topographic conditions that allowed groundwater table to rise and inundate bottoms of trenches; *waste stability* — wastes not properly segregated nor drained.

Despite all these difficulties, "no radiological releases of health and safety significance have occurred."

Most studies on discharges from low-level burial grounds have concentrated on the movement by groundwater. However, it has also been pointed out that radionuclide doses from low-level radioactive waste burial sites caused by burrowing animals and translocation by plants are of the same order of magnitude as those of the more commonly evaluated human intrusion scenario.[30]

In addition, to complete the mass balance of radionuclides leaving the disposal trenches, the movement through the unsaturated zone by gaseous transport is being investigated at the site in Sheffield, Illinois, by the USGS.[31] The spatial and temporal movements of radon-222, tritiated water, and C-14 dioxide are being studied.

More complete understanding of the movement of leachates from trenches is being obtained by studies at the site in Maxey Flats, Kentucky.[32] Both plutonium and cobalt have been chelated by EDTA. Though a number of nuclides—H-3, Pu-238, Pu-239, Pu-240, Co-60, Cs-137, and Sr-90—have moved a few feet to 50 ft on site, concern is expressed about potentially greater and swifter movement, because plutonium and cobalt appear to exist as anionic species with organic properties. Hydrophobic organic compounds, including barbiturates, are in the leachate at low ppb concentrations.

COMPARISON WITH DISPOSAL OF LOW-LEVEL WASTES IN OTHER COUNTRIES

The practices for the disposal of low-level wastes in the United States and in the United Kingdom are almost unique in that these two countries allow direct near-surface disposal of wastes without any treatment of the wastes. English practice, as exemplified at Drigg, in Cumbria, is shown in Figure 1.[33] U.S. practice is neater, to conserve space, but leaves the wastes as open to leaching as does the English system. The French practice near-surface disposal in tumuli and monoliths, but only after extensive treatment. A schematic of the disposal site is shown in Figure 2, and photos of both tumuli and monoliths in Figures 3–7.[34] The only leachate to date from La Manche has been tritium. It is believed that it was due to leachate developed before the tumuli were covered.

Sweden, Switzerland, and the Federal Republic of Germany have chosen to place their low-level wastes at greater depths. In Germany, the wastes are placed in the disused Konrad iron ore mine at depths of 800 to 1,300 m. The wastes will be placed in new tunnels and drifts drilled for that purpose. The capacity of the repository is estimated to be at least 500,000 m³.[35]

The Swiss, as part of Project Gewahr, have chosen, for the present, to put the low-level waste together with the intermediate-level waste in a marl formation 750 to 1,300 m below ground. (They have reserved the right to construct, later, a facility for low-level waste only.) The Swedes are constructing a low- and intermediate-level waste facility in granite about 50 m below the seabed at the Forsmark Nuclear Power Station. The capacity will be 90,000 m³, which should be large enough to contain all low- and intermediate-level wastes, other than decommissioning wastes, produced until the year 2010. All low-level waste has a half-life of 30 yr or less. All intermediate-level waste will be encapsulated in concrete or bitumen.[36] A diagram of the the facility under construction appears in Figure 8.[37]

Figure 1. Filling and monitoring of near-surface disposal trench at Drigg, Cumbria, England. (Source: House of Commons.[33])

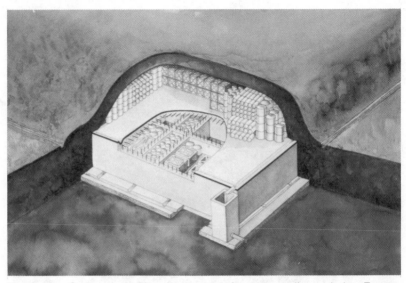

Figure 2. Centre de la Manche, near-surface waste disposal site, France. The structure is designed for comprehensive protection of the waste containers against external forces; should this protection fail, it is also designed to reduce adverse consequences of accidents. This dual role is accomplished through a series of barriers: sand, clay, and gravel in the tumulus; the concrete walls of the monoliths; the walls of the individual containers; and the covering of the waste itself. The structure is earthquake resistant and impervious to rainwater and groundwater. At its base there is a network of channels, accessible by shafts, for monitoring the watertightness of the tumulus, and another network of channels for catching rainwater. (Source: ANDRA.[34])

Figure 3. Tumulus, Centre de la Manche. A frame for the tumulus is formed by placing some units on the periphery of the storage floor and other units in rows in the middle of the floor. (Source: ANDRA.[34])

Figure 4. Tumulus, Centre de la Manche. A 3,000-m² storage floor can hold approximately 10,000 m³ of waste containers. (Source: ANDRA.[34])

Figure 5. Monolith, Centre de la Manche. A control board guides the placement of coffering in the trench. Arranging metal gridwork on the bottom and along the sides of the enclosure guarantees structural integrity. The containers are then lowered into the enclosure with a crane. (Source: ANDRA.[34])

Figure 6. Monolith, Centre de la Manche. The waste containers are packed in layers. Concrete poured over each layer covers the containers completely. (Source: ANDRA.[34])

Figure 7. Monolith construction, Centre de la Manche. A large excavation is made first and its bottom is covered with a layer of concrete. Around this concrete floor, a channel and intermediate trenches are dug to collect seepage and rainwater during construction. (Source: ANDRA.[34])

A   Silo repository
B   Concrete tank repository
C   Rock vault for low-level waste
D   Rock vault for intermediate-level
     waste
E   Operating tunnel
F   Construction tunnel

Figure 8. Diagram, low- and intermediate-level waste facility, Forsmark Nuclear Power Station, Sweden. The illustration shows, at the left, a rock vault for low-level waste in which deposition is in progress and, at the right, a rock vault for intermediate-level waste, also in the deposition phase. The low-level waste is handled by an ordinary forklift truck. After completed deposition, the entrances to the chamber are sealed with concrete plugs. The intermediate-level waste is placed in a rock vault partitioned by concrete walls. The waste is handled by a remotely controlled overhead travelling crane. After deposition, the metal drums and metal cases are grouted with concrete. When the facility is sealed, supplementary grouting and backfilling are carried out. The entrances to the

CONCLUSIONS

Because low-level radioactive wastes are defined by law, some wastes are included that would not necessarily be there if disposal options were based upon the risks associated with the material. We pay an exorbitant price for this luxury. Between 1985 and 2000, it has been projected that 65% (80 × 10⁶ ft³) of the nation's total low-level radwaste volume will be produced by nuclear fuel cycle sources. The remaining 35% (45 × 10⁶ ft³) will be generated by non-fuel cycle sources, such as hospitals and research laboratories. Most of this waste will contain radionuclides (primarily Co-60, Cs-137, H-3, C-14, and I-131) in low concentrations. Most wastes will be packaged in 55-gal drums or wooden boxes, with 50% of the packages having surface radiation dosages < 2 mR/hr.

Although there has been measurable movement of radionuclides onsite at the three former commercial low-level radioactive waste disposal sites, no significant (in terms of dose) movement offsite has occurred. No significant movement of radionuclides has been detected at the one site still operating in a humid region of the country nor at the two sites operating in arid regions.

Because of the complex hydrogeology and geochemical reactions taking place and the lack of a comprehensive mass balance study at any of the sites, no definitive conclusions on the mechanisms and rates of transport can be made about the low levels of movement that have taken place. Because degradation and solidification are accelerated in the presence of water, many of major changes to be expected have already taken place, so that catastrophic or major releases should not be expected at some future time. However, the exact rate of future movements cannot be accurately forecast.

Based upon public perception in this country and abroad, it is clear that the most efficient technical solution or the most efficient economic solution, meeting all of the health and risk guidelines, will not be accepted by the public. Therefore, a socially acceptable solution is required. The time and cost of obtaining approval of the most efficient technical or economic disposal method will likely be higher than for the socially acceptable one. In economists' terms, the transaction costs are so high that a solution with low transaction costs should be sought. Based on experience here and abroad, a higher degree of treatment or disposal at greater depth will be required.

APPENDIX

1 rad = 0.01 J kg^{-1} = 0.01 Gy (Gray)
1 curie = 3.7 × 10^{10} s^{-1} = 3.7 × 10^{10} Bq (Becquerel)
1 rem = 0.01 Sv (Sievert)

REFERENCES

1. 1970 Standardization of Radioactive Waste Categories Technical Report, International Atomic Energy Agency, Series No. 101.
2. Low-Level Radioactive Waste Policy Act, P.L. 96–573, December 22, 1980.
3. "Licensing Requirements for Land Disposal of Radioactive Waste," 10 CFR 61.
4. Nuclear Waste Policy Act of 1982, P.L. 95–425, January 7, 1983.
5. Atomic Energy Act of 1954, 42 U.S.C. 2014(e)(27).
6. 40 CFR 191, August 15, 1985, 191.02(i).
7. 10 CFR 61, 61.7(a)(l) and (2).
8. 10 CFR 61, 61.55(a)(3).
9. 10 CFR 61, 61.55(a)(4).
10. 10 CFR 61, 61.7(b)(2).
11. 10 CFR 61, 61.55 (a)(2)(iv).
12. Gruhlke, J. M., and Galpin, F. L., "Investigations of Potential 'Below Regulatory Concern' (De Minimis) Wastes," in *Proceedings of the Fifth Annual Participant's Information Meeting, DOE Low-Level Waste Management Program* (Idaho Falls, ID: National Low-Level Radioactive Waste Management Program, 1983), pp. 359–380.
13. U.S. National Research Council Value/Impact Statement of Amendments to 10 CFR 20 for Disposal of Biomedical Wastes, March 2, 1981.
14. "Radiation Protection Principles for the Disposal of Solid Radioactive Waste" (Publication 46). International Commission on Radiological Protection, July 1985.
15. Final Environmental Impact Statement for 10 CFR 61.
16. Parker, F. L., and Taboas, A., "Transuranic Waste Disposal," in *Proceedings, Hazardous Materials Management Conference and Exhibition* (Philadelphia, PA, 1984).
17. Stone, J. A., et al., "Radionuclide Migration Studies at the Savannah River Plant Humid Shallow Land Burial Site for Low-Level Radioactive Waste," in *Proceedings of the Sixth Annual Participant's Information Meeting, DOE Low-Level Radioactive Waste Management Program* (Idaho Falls, ID: National Low-Level Radioactive Waste Management Program, 1984), pp. 119–129.
18. Herrington, J. S. (U.S. secretary of energy), personal communication, April 30, 1985.
19. Resource Conservation and Recovery Act of 1976, P.L. 94–580.
20. "Identification and Listing of Hazardous Wastes," 40 CFR 261.
21. Jacobs, D. G., and J. W. Lynch, "Comparison of Regulations for Low-Level Waste Disposal and Hazardous Waste Disposal" (Oak Ridge, TN: Martin Marietta Energy Systems, DOE/HWP-8, 1985).
22. Resource Conservation and Recovery Act Amendments of 1984, P.L. 98–616.
23. "Risk, Science and Democracy," *Issues in Sci. and Tech.* 1(3):19–38 (1985).
24. Sinclair, W., National Council on Radiation Protection, personal communication to Cochran, R., U.S. Department of Energy, April 11, 1986.

25. Sauer, R. E., "A Commercial Regional Incinerator Facility for Treatment of Low-Level Radioactive Waste," in *Proceedings of the Sixth Annual Participant's Information Meeting* (Idaho Falls, ID: National Low-Level Radioactive Waste Management Program, 1984), pp. 377–388.

26. "Draft Environmental Impact Statement—Central State Disposal Facility for Low-Level Radioactive Waste—Oak Ridge Reservation, Oak Ridge, TN" (E/EIS-0110-D). Washington, DC: U.S. Department of Energy, 1984.

27. Fischer, J. N., "Low-Level Radioactive Waste Program of the U.S. Geological Survey—In Transition," in *Proceedings of the Fifth Annual Participant's Information Meeting* (Idaho Falls, ID: National Low-Level Radioactive Waste Management Program, 1983), pp. 52–61.

28. Fischer, J. N., "U.S. Geological Survey Studies of Commercial Low-Level Radioactive Waste Disposal Sites—A Survey of Results," in *Proceedings of the Fifth Annual Participant's Information Meeting* (Idaho Falls, ID: National Low-Level Radioactive Waste Management Program, 1983), pp. 650–658.

29. Oztunali, O, "Lessons Learned in Disposal Design," in *Proceedings of a Symposium on Low-Level Radioactive Waste Disposal: Facility Design, Construction, and Operating Practices* (NUREG/CP-0028), Volume 3 (Washington, DC: U.S. Nuclear Regulatory Commission, 1983).

30. Cadwell, L. L., et al., "Evaluating Biological Transport of Radionuclides at Commercial Low-level Waste Burial Sites," in Proceedings of the Fifth Annual Participant's Information Meeting (Idaho Falls, ID: National Low-Level Radioactive Waste Management Program, 1983), pp. 560–576.

31. Striegl, R. G., "Methods for Determining the Transport of Radioactive Gases in the Unsaturated Zone," in *Proceedings of the Sixth Annual Participant's Information Meeting* (Idaho Falls, ID: National Low-Level Radioactive Waste Management Program, 1984), pp. 579–584.

32. Toste, A. P., et al., "Chemical Characteristics, Migration and Fate of Radionuclides at Commercial Shallow Land Burial Sites," in *Proceedings of the Sixth Annual Participant's Information Meeting* (Idaho Falls, ID: National Low-Level Radioactive Waste Management Program, 1984), pp. 85–99.

33. First Report from the Environmental Committee on Radioactive Waste to the House of Commons Session, 1985–1986, Volume 1 (Report together with the Proceedings of the Committee Relating to the Report, January 28, 1986.)

34. "Le Centre de la Manche," Commissariat à l'Énergie atomique —Agencie Nationale pour la Gestion des Déchets radioactifs (ANDRA), 1982.

35. Parker, F. L., et al., "Technical and Sociopolitical Considerations in Selected Radioactive Waste Policy Issues" (Stockholm, Sweden: The Beijer Institute, Royal Swedish Academy of Sciences, in press).

36. SKI—Swedish Nuclear Power Inspectorate, "Licensing of Final Repository for Reactor Waste" (Technical Report SKI 84:2) (Stockholm, Sweden: SKI, 1984).

37. Swedish Nuclear Fuel and Waste Management Company, "Final Repository for Reactor Waste—SFR Information Booklet" (Stockholm, Sweden, 1986).

CHAPTER 5

Low-Level Radioactive Waste at University Medical Centers

Michael J. Welch, Barry A. Siegel, John O. Eichling, and Carla J. Mathias

Low-level radioactive waste generated at university medical centers arises from activities in both the clinical practice of medicine and biomedical research.[1,2] The clinical practice area includes nuclear medicine, in vitro radioassays, and radiation therapy. Biomedical research involves many types of in vitro studies and animal studies performed with small amounts of radioactivity, predominantly C-14 and tritium. In this chapter, we will discuss briefly these areas of use and the types of waste they generate and, using the Washington University Medical Center in St. Louis, Missouri, as an example, will discuss the impact of recent regulations upon the handling of low-level radioactive waste.[3-5]

SOURCES OF WASTE

Clinical nuclear medicine is now a required service for full accreditation of hospitals. Nuclear medicine studies are performed on one in three patients admitted to hospital in the United States; approximately 12 million nuclear medicine studies are performed annually in the United States (with nearly 21,000 per year done at the three hospitals that make up the Washington University Medical Center and operate under a single institutional broad license from the Nuclear Regulatory Commission). The most common types of studies and the radionuclides most commonly used (and their half-lives) are given in Table 1. In the vast majority of diagnostic examinations performed by clinical nuclear medicine departments, a radiolabeled compound (radiopharmaceutical) is administered to a patient (most commonly by intravenous injection) and, at varying intervals thereafter, images of the distribution of the radioactiv-

Low-Level Radioactive Waste Regulation: Science, Politics, and Fear, Michael E. Burns, Ed., © 1988 Lewis Publishers, Inc., Chelsea, Michigan – Printed in USA.

Table 1.   Radionuclides Used in Clinical Nuclear Medicine

Radionuclide	Half-life	Diagnostic Procedure
^{99}Mo/^{99m}Tc	67 hrs/6 hrs	Thyroid, brain, bone, kidney, gastrointestinal, liver, heart imaging > 70%
^{123}I	13 hrs	Thyroid, brain imaging
^{111}In	67 hrs	Infection, spinal fluid, radiolabeled antibody imaging
$^{/201}$Tl	73 hrs	Heart, parathyroid imaging
^{67}Ga	78 hrs	Tumor, infection imaging
^{131}I	8 days	Kidney, thyroid, radiolabeled antibody imaging
^{51}Cr	28 days	Blood volume and red cell survival
^{169}Yb	32 days	Spinal fluid imaging
^{125}I	60 days	Blood volume
^{57}Co	270 days	Vitamin B$_{12}$ absorption

Waste—Spent generators, unused radiopharmaceuticals, syringes, glassware, gloves, etc.

ity within the patient are obtained by external detection of the emitted gamma or characteristic X-ray photons with an instrument known as a scintillation camera. The resultant images are referred to as scintigrams. Scintigraphic studies provide a wide variety of clinically important diagnostic information, which reflect normal physiologic function vs pathophysiologic aberrations in function, as well as normal vs abnormal anatomy. For example, an image of the lungs obtained after injection of radiolabeled particles that temporarily occlude a small fraction of the pulmonary capillaries reflects the gross anatomic structure of the lungs, but more importantly indicates the relative regional distribution of blood flow within the lungs. Hence, this test is used to diagnose a common disorder, pulmonary embolism, in which thrombi initially formed in the peripheral veins have dislodged and been carried by the blood to the pulmonary vessels they obstruct. In such cases, the pulmonary perfusion scintigrams demonstrate focal areas of decreased or absent perfusion corresponding to the portion of the lungs not receiving normal blood flow. Another common test, bone scintigraphy, employs a radioactive tracer that localizes in normal bone but accumulates to a greater degree at most sites of abnormal bone turnover; this test provides a highly sensitive means to detect skeletal diseases due to trauma, infection, tumor, and other conditions at a time when conventional diagnostic roentgenograms are normal. Similar diagnostic nuclear medicine examinations have been developed for evaluation of the anatomy and function of most organs in the body. The waste generated by clinical nuclear medicine activities includes spent generators, expired vials of radiopharmaceuticals, and contaminated syringes, glassware, gloves, absorbent pads, etc. It should be noted that about 70% of clinical nuclear medicine studies are carried out with the use of the molybdenum-99/technetium-99m generator; the half-life of the parent radionuclide (Mo-99) is relatively short ($t_{1/2}$ = 67 h), and the daughter, Tc-99m ($t_{1/2}$ = 6.02 hrs), is utilized to produce a variety of routine radiopharmaceuticals. The majority of the low-level radioactive waste generated by clinical nuclear medicine, then, is Tc-99m.[6] Essentially all waste contaminated with Tc-99m may be held for decay and then disposed of along with nonradioactive trash (with appropriate segregation of biohazardous materials, e.g., used needles). Only a few of the nuclides commonly used in nuclear medicine have half-lives longer than 28 days; these include I-125, Co-57, and Yb-169, all of which are used in very small quantities. All of the radionuclides employed in nuclear medicine are, however, prepared and processed by radiopharmaceutical manufacturers throughout the country. The separation of several of these radiopharmaceuticals, in particular molybdenum-99 or iodine-131, which are products of uranium fission, results in large amounts of radioactive waste. The largest disposal problem relating to the clinical practice of nuclear medicine is, in fact, faced by the producers of radiopharmaceuticals rather than by the users themselves.[6]

In vitro radioassays are essential clinical and research techniques for measurement of hormones and other biological substances present in plasma or other materials at very low concentration (ng/mL). Over 100 million of these sensitive assays are carried out annually in the United States (with over 150,000 annually performed at the Washington University Medical Center). The major radionuclide utilized for radioassay is iodine-125, which has a half-life of 60 days. The waste generated in these procedures involves reagents, gloves, test tubes, and other disposable laboratory supplies.

The third clinical area is radiation therapy, which involves the use of high intensity radiation sources in the primary or palliative treatment of cancer. Approximately 400,000 patients per year undergo radiation therapy in the United States (with 2,300 treated at Washington University Medical Center). Although much radiation therapy is performed with linear accelerators and thus does not generate radioactive waste, a variety of important therapeutic procedures still involves both sealed and unsealed radionuclides. The radionuclides used in these applications include yttrium-90, phosphorus-32, cobalt-60, iodine-125 and iodine-131, cesium-137, and iridium-192; these isotopes have a range of half-lives from 2.6 days to 30 years. The radionuclides are in various forms for use as therapy sources; radiopharmaceuticals, which are administered orally, intravenously, or by intracavitary injection; sealed sources, which are temporarily implanted into patients; sealed sources, e.g., small seeds or wires, which are permanently implanted directly in the area to be treated; and sealed sources used for external beam therapy. The major type of waste generated from radiation therapy, then, is the spent sources; these tend to have the greatest amount of radioactivity per unit mass or volume of all wastes generated at medical institutions.

The second area that generates low-level radioactive waste in medical centers is biomedical research. Radionuclidic tracers are currently used in most types of modern biomedical research.[7] A survey of journals in the areas of biochemistry, immunology, endocrinology, and metabolism shows that almost 50% of all modern biomedical research involves the use of radioactive tracers. At the Washington University Medical Center, over 360 individual faculty members working in over 600 individual laboratories regularly use radioactive materials in their research. Over 40% of all biomedical research grants funded at the present time involve the use of some radioactive materials, most commonly tritium, carbon-14, phosphorus-32, sulfur-35, and iodine-125. The major radioactive waste produced is the small quantity of activity contained in liquid scintillation vials, over a million of which are generated each year at the Washington University Medical Center. Other forms of waste are unused reagents, contaminated laboratory materials, and animal carcasses. It should be noted that as part of the preclinical evaluation of nonradioactive drugs, extensive in vivo studies of the kinetics and metabolism of the drugs are required prior to submission of a new drug application (NDA) to the Food and Drug Administration. The majority of these studies are carried out with radiolabeled counterparts of the drugs; thereby, low-level radioactive waste is produced in the preclinical development of essentially all ethical drugs.

SAFETY PROGRAM

The overall radiation safety program at Washington University Medical Center involves: 2,200 monitored workers, the 360 faculty members mentioned above, in whom is vested the supervisory responsibility for the use of radionuclides in individual research laboratories; seven diagnosis and therapy groups; and two medical cyclotrons used to produce positron-emitting radionuclides for research and diagnostic studies. The low-level radioactive waste disposal for a recent year is shown in Table 2. Table 3 shows the sources of this low-level radioactive waste and an estimate of the relative disposal costs from the various areas. Tables 4 and 5 show the types and forms of material disposed of, and the costs. It should be noted that the majority of the volume and the costs are associated with the biomedical research application of radioisotopes, rather than with the diagnostic or therapeutic applications. As previously discussed, the bulk of the waste related to clinical activities is generated by the manufacturers of the isotopes. At Washington University, 88% of the total volume of low-level radioactive wastes (containing 99% of total activity) was disposed of by burial at one of the three low-level radioactive waste sites open at that time, while 12% of the volume (containing 1% of the activity) was disposed of by incineration.

The primary approach to reduction of the amount of material sent for

Table 2. Annual Low-Level Radioactive Waste Disposal, Washington University, 1984

Radionuclide	$T_{1/2}$	% Total Activity
^3H	12.2 yr	39.6%
^{125}I	60 d	25.3%
^{32}P	14 d	16.5%
^{35}S	88 d	15.0%
^{51}Cr	28 d	1.4%
^{14}C	5730 yr	1.0%
^{45}Ca	165 d	0.5%
^{131}I	8 d	0.2%

Total Activity 94. Ci
$T_{1/2} < 90$ d—58.5%
$T_{1/2} < 12.2$ yr—99%

Table 3. Sources of Low-Level Radioactive Waste, Washington University, 1984

	Annual Volume	Disposal Cost
Clinical Nuclear Medicine	20 ft^3	$500
Radioassay	500 ft^3	$40,000
Radiation Therapy	3–4 ft^3	$100
Research	8400 ft^3	$306,000

Table 4. Low-Level Radioactive Waste Disposal Costs, Washington University, 1984

Category	Method	Unit Cost	Annual Cost
Dry Solids	Burial	$1.00/lb	$ 59,000
	Incineration	$0.80/lb	1,000
Absorbed Liquids	Burial	$40.00/gal	228,000
Scintillation Vials and Contents	Burial	10.5 cents/vial	63,000
	Incineration	3.5 cents/vial	16,000
Animal Carcasses	Burial	$2.25/lb	14,000
			$ 381,000
		+ Labor Cost	28,000
		Total Annual Cost	$ 409,000

Table 5. Forms of Low-Level Radioactive Waste, Washington University, 1984

	Annual Quantity		
	Weight (lbs) (%)	Volume (ft^3) (%)	Activity (Ci) (%)
Dry Solid	61,000 (28%)	2,200 (25%)	2.80 (30%)
Absorbed Liquid	71,000 (32%)	3,100 (35%)	5.80 (62%)
Scintillation Vials and Contents	78,000 (35%)	3,100 (35%)	0.31 (3%)
Animal Carcasses	11,000 (5%)	500 (6%)	0.47 (5%)
Total	221,000 (100%)	8,900 (100%)	9.38 (100%)

burial and incineration is segregation of the various types of waste. A breakdown of the types of waste produced and disposal practice is given in Tables 6 and 7. A significant proportion of the waste being produced contains radionuclides with half-lives greater than 90 days or relatively small amounts of radioactivity in scintillation vials also containing organic solvents. The bulk liquid, which has radioactive material with half-lives of less than 90 days, can be stored to allow for decay and, after decay, can be disposed by discharge into the sanitary sewage system.

Table 6.   LLRW Waste Quantities, Washington Medical Center, 1984

Dry with half-life < 90 days	4200 ft^3
Dry with half-life > 90 days	700 ft^3
Scintillation vials with organic solvents	1.7 million vials
Aqueous liquid with half-life < 90 days	4800 gal
Organic liquid with half-life > 90 days	700 gal
Organic liquid with half-life < 90 days	500 gal
Animal carcasses with half-life < 90 days	1400 ft^3
Animal carcasses with half-life > 90 days	200 ft^3
Total Weight:	300,000 lbs

Small amounts of deregulated material containing tritium or C-14 in scintillation vials can be shipped for incineration at approved sites.[4] Other short-lived organic material also can be shipped for incineration. Therefore, the only material that needs to be shipped for burial is material with half-lives greater than 90 days, including solid, aqueous, or organic material, with high levels of radioactivity. Segregation of the radioactive waste into these classes reduces the total amount of material shipped for burial from major medical centers, and could have significantly reduced the amount that was shipped over the past few years (Table 7). This will ultimately increase the useful life of the burial disposal sites for low-level radioactive waste.

Over the past 15 years, the Radiation Safety Office of Washington University Medical Center has increased from one part-time faculty member, one full-time employee, and a part-time secretary to a group of 12 employees with an annual budget of $650,000; the operation now has a significant self-storage capacity. The storage capability is needed for the decay of the short-lived radioisotopes and also for emergency storage

Table 7.   Current Low-Level Radioactive Waste Disposal Practices, Washington University (Effective October 1985)

Waste Type	Category	Disposal Method
Bulk Liquid	Aqueous < 90 day half-life	Hold for decay and local drain release
	Aqueous > 90 day half-life	Ship absorbed for burial via Adco
	Organic deregulated (^{3}H and/or ^{14}C < 0.05 μCi/g)	Ship in bulk to Quadrex via Adco for incineration
Bulk Liquid(SV)[a]	Organic containing ^{125}I, ^{51}Cr, ^{59}Fe, ^{35}S, ^{32}P, ^{45}Ca, ^{22}Na, ^{57}Co, ^{86}Rb, ^{67}Ga, ^{65}Zn, ^{111}In, ^{36}Cl, ^{203}Hg, ^{33}P, ^{131}I, ^{75}Se, ^{68}Ge, ^{109}Cd, ^{141}Ce, ^{46}Sc, ^{64}Cu, ^{195}Au, ^{99}Tc, ^{153}Gd, ^{113}Sn, ^{119}Sn and ^{3}H and/or ^{14}C in concentrations too high to be "deregulated"	Ship in bulk form to Quadrex via Adco for incineration
Bulk Liquid(SV)[a]	Organic not listed above	No convenient disposal route
Dry	< 90 day half-life	Hold for decay and compact and ship for burial via Adco
	> 90 day half-life	Hold for decay and compact and ship for burial via Adco
Animal Carcasses	Deregulated (^{3}H and/or ^{14}C < 0.05 μCi/g)	Ship for burial via Adco
	< 90 day half-life	Hold for decay and local incineration
	> 90 day half-life	Ship for burial via Adco

[a]Including scintillation vials with liquid contents in these categories.

to enable the medical center to store its long-lived waste for a total time of ten years (in the event of shut-down of low-level waste storage sites).[3-5] The total storage capacity will accommodate 2000 drums (55 gallons), of which 1000 are for the decay of short-lived radioisotopes and 1000 are held in reserve for emergency storage of long-lived radioactive waste. At present a maximum of 100 drums of long-lived radioactive waste are being generated per year; thus the medical center has an emergency storage capacity sufficient to store waste for ten years. An alternative to long-term storage is onsite incineration, and this is certainly a consideration at many large research institutions.

Major medical centers will continue to produce low-level radioactive waste as a by-product of essential activities in four areas: diagnosis, radioassay, therapy, and research. The amount of such waste is likely to grow in volume, and the costs of its disposal will likely increase in the current regulatory environment. Extensive planning for the separation and proper disposal of radioactive waste can yield an appropriate compromise for immediate and safe handling of low-level radioactive waste produced by these essential clinical and biomedical research activities.

REFERENCES

1. Vance, J.N., "Processing of Low-Level Waste," in Koval, T.M. Ed., *Radioactive Waste* (Bethesda, MD: National Council on Radiation Protection and Measurements, 1986), pp. 38–53.
2. Carter, N.W., and Stone, D.C., "Quantities and Sources of Radioactive Waste," in Koval, T.M. Ed., *Radioactive Waste* (Bethesda, MD: National Council on Radiation Protection and Measurements, 1986), pp. 5–30.
3. "Compact Bills in U.S. Move One Step Ahead, Policy Changes in Canada and France," *J. Nucl. Med.* 26:845–846 (1985).
4. Frohman, J.E., "EPA Forces Low-Level Disposal Site to Stop Accepting Scintillation Vials: Brookhaven Studies Chemical Characteristics of Radioactive Waste," *J. Nucl. Med.* 26:1366–1367 (1985).
5. Brill, D.R.: "Low-Level Radioactive Waste Compacts: One Year and Counting," *J. Nucl. Med.* 26:2–6 (1985).
6. Anderson, R.C.; Beck, T.J.; Cooley, L.R.; and Strauss, C.S., "Institutional Radioactive Wastes," Report NUREG/CR 0028, U.S. Nuclear Regulatory Commission (March 1980).
7. Yalow, R.S., "Disposal of Low-Level Radioactive Waste: Perspective of the Biomedical Community," in Koval, T.M. Ed., *Radioactive Waste* (Bethesda, MD: National Council on Radiation Protection and Measurements, 1986), pp. 59–64.

CHAPTER 6

Low-Level Radioactive Wastes in the Nuclear Power Industry

Robert A. Shaw

INTRODUCTION

The first nuclear reactor was operated briefly by Enrico Fermi and his associates in Chicago on December 2, 1942, less than four years after the discovery of nuclear fission. For the next 15 years, nuclear power plants were developed primarily for military and defense purposes. Some were experimental, designed to determine more of the important features of such power plants; others were production facilities for the production of plutonium used in the manufacture of nuclear weapons. In 1957 the first nuclear power plant for producing electricity in the United States was developed at Shippingport, Pennsylvania. Even this was not yet a commercial facility: it was associated with the naval reactor portion of the nuclear program.

Starting in 1959, commercial nuclear power plants, exclusively for generating electricity, were brought into operation. At the end of 1986, approximately 100 nuclear power plants were in commercial operation in the United States. The number is approaching 400 worldwide. A few countries now produce more than half of their electricity from nuclear power plants.

There are also nuclear power plants used for military and defense purposes. Since waste processing procedures and rules are frequently different for these plants, this chapter will consider commercial nuclear power plants only. Radioactivity is produced directly in nuclear power plants as a result of the energy-producing fission process. When the uranium nucleus splits, the two resulting particles or fission fragments are radioactive—unstable in the nucleus—and will decay, emitting parti-

Low-Level Radioactive Waste Regulation: Science, Politics, and Fear, Michael E. Burns, Ed., © 1988 Lewis Publishers, Inc., Chelsea, Michigan—Printed in USA.

cles or rays, in order to approach a nuclear equilibrium. Common among such fission fragments are radioisotopes of strontium, xenon, cesium, krypton, and many other elements.

Radioactivity is formed indirectly in nuclear power plants through the production of neutrons in fission. These neutrons bombard other materials in the area, producing more radioactive nuclei that will similarly decay. Typical among these indirectly formed radioisotopes are those of cobalt, iron, manganese, and chromium.

The fragments that are formed directly from the fission process are generally contained within the fuel cladding. This fuel material, once it has been used within a nuclear power plant, becomes highly radioactive. The waste resulting from the disposal of such fuel is classified as high-level radioactive waste (HLRW). Such fuel may be processed in order to separate the deleterious fission fragments from the remaining uranium. Although some countries are reprocessing their fuel, the United States is not. Consequently, the HLRW from the commercial nuclear power industry in the United States is in the form of used or spent fuel.

Except for the fuel, the radioactive wastes generated within the nuclear power plant are termed low-level radioactive wastes (LLRW). The primary source of such LLRW is the contaminants contained in the cooling water that is used to generate the steam in a nuclear power plant. These are generally corrosion products that are released from the surfaces within a nuclear power plant. Most such surfaces are stainless steel, although there are Inconel-600 and cobalt alloys present that can also make significant contributions. The release of corrosion products from these surfaces and the transport of these corrosion products to the reactor core results in the generation of radioisotopes.

A portion of these corrosion product radioisotopes will be released from core deposits to the reactor water coolant. Some will redeposit on surfaces within the plant that are remote from the core, and a radiation field will develop around these areas of deposition.

Small amounts of fission products may also be found in reactor coolant water. These result from stray surface contamination of fuel cladding by uranium, or from pinhole leaks in fuel cladding that permit very limited release of fission products to the coolant water.

The purification and filtration of the reactor coolant water removes these contaminants. These contaminants will include both nonradioactive and radioactive species. These filtration and purification processes are a significant source of LLRW—e.g., spent filters—in nuclear power plants.

Classification of waste as HLRW or LLRW is based on the source of the waste, rather than radiation levels. Although in most cases high-level waste will be more radioactive than low-level waste, some LLRW components, such as in core hardware, are highly radioactive.

The remainder of this chapter will include discussions of LLRW sources, monitoring, processing, and volume reduction and disposal. The Electric Power Research Institute states,

> the basic function of the radwaste (radioactive waste) system is to:
> 1. Minimize the release of gaseous radwaste to the environs through delay and filtering;
> 2. Minimize the release of liquid radwaste to the environs by purifying or reclaiming plant waste water; and
> 3. Minimize the impact of shallow land disposal by producing a solid waste product which is in compliance with federal criteria.[1]

RADWASTE SOURCES

Power Plant Coolant Circuits

All commercial nuclear power plants in the United States are either Pressurized Water Reactor (PWR) or Boiling Water Reactor (BWR) types, except for one high-temperature gas-cooled reactor. Despite the basic differences in the nature of these two types of power plants, most of the sources, treatment, and processing of LLRW are sufficiently similar that both will be handled in the same fashion.

Low-level wastes produced in these nuclear power plants are "wet" or "dry." "Wet" wastes are by-products of relatively standard water purification and treatment systems within the plant. "Dry" wastes are discarded equipment, hardware, and trash that may have radioactivity associated with them. These two waste types are described further below.

Water is the coolant for nuclear power plants. In the BWR, the water is passed through the reactor core and boiled, and the steam is sent directly to the turbines. The steam expands through the turbine and is condensed and returned to the core to be boiled again. In the PWR, the water is passed through the reactor core under sufficient pressure to prevent boiling, and then to a heat exchanger called a steam generator. In the secondary side of the steam generator, water is converted to steam, which then turns the turbine. The steam is condensed and returned to be heated again in a closed cycle. The steam generator tubes supply the surfaces that permit the transfer of heat from the primary cooling water, which cools the fuel, to the secondary cooling system, and the tubes maintain a physical barrier between the primary and secondary system.

To maintain impurities at sufficiently low concentrations and to maintain desired coolant chemistry, a portion of the water passing through the reactor core is treated in a water purification system. This purification system is generally referred to as the letdown system in the PWR and as the reactor water cleanup system in the BWR. Both systems use relatively standard water treatment processes of filtration and ion exchange to remove both soluble and insoluble radioisotopes and impurities present in the reactor coolant. When the ion exchange resins and filters are spent, they are discharged so that the desired purification characteristics of the system are maintained. The discharged resin and filters then become one of the significant sources of LLRW.

The BWR, additionally, has resins that polish and purify the condensate coming from the turbine. This condensate will be very mildly radioactive; so, consequently, will be the condensate polishing resin. Similarly, some PWRs have condensate polishing resins in the secondary cooling circuit. In any PWR that has had some steam generator leaks, so that small amounts of radioactive reactor coolant have escaped into the secondary circuit there will then be some radioactive constituents in the secondary side of the coolant loop. Such leaks are required by NRC limits to be very small, on the order of 1/3 gallon per minute. Plants repair or plug leaking tubes regularly to prevent such leakage. Nonetheless, condensate polishing resins in these plants will be radioactive and must be discharged as LLRW.

Leakage of small quantities of coolant from valve stems, pump seals, etc., from systems within the plant are collected along with floor washdown water for treatment to remove impurities and radioactivity. The drains from chemical laboratories will contain radioactivity from the required testing of reactor coolant samples and other such materials in the laboratory. Laundry drains will contain the detergent solutions used to clean protective clothing worn by the workers in radioactive environments. All of these liquid waste streams are processed in order to concentrate the radioisotopes and produce water of sufficient purity for recycling back into use at the plant. Table 1 presents a compilation of liquid waste sources.

Waste Products

Several processes are used by nuclear power plants to concentrate liquid waste streams. Some plants use an evaporator to concentrate the waste stream. Others use a combination of a filter and ion exchange resin similar to the one already mentioned. Membrane processes and other concentration processes have been considered, and in some cases used, to concentrate the constituents in these liquid streams. The end product in

Table 1.   Reactor Liquid Wastes

| Waste Type | Waste Source | | Disposal Method |
	PWR	BWR	
High purity	Reactor coolant samples, bleeds, and leaks	Reactor water leads, samples, condensate leaks	Recycle
Low purity	Leakage onto floors, floor washdowns, etc.	Same as PWR	Recycle or discharge
Chemical waste	Laboratory wastes, tool decontamination and regeneration solutions	Same as PWR	Discharge or solidify
Detergent	Laundry, tool decontamination	Same as PWR	Recycle or discharge

all cases is a liquid stream more highly concentrated in particulates, and in some cases it is as thick as sludge.

Another constituent of radioactive waste is commonly referred to as dry active waste (DAW). Dry active waste generally contains a wide range of substances, usually materials with extremely low radioactive content: covers, miscellaneous paper, construction materials, material which has been used in the reactor environment, etc. In many cases, nuclear power plants reduce the volume of DAW by compacting the compressible portions of these wastes, particularly plastics.

Figure 1 is a schematic of a radwaste treatment system typical of BWRs. It traces waste from its sources (on the left side of the form) through processing (in the center) to its final packaged form (on the far right).

Waste volume sources and radiological characteristics of waste during 1978–81 were determined in an industry survey.[2] These results, together with more recent data, are presented in Figures 2 and 3 for PWRs and BWRs, respectively. They show the trend in waste volumes shipped from these units, for both average and median plants.

As a result of federal legislation passed in 1985, allocations were established for the volume of radioactive wastes that could be shipped from any power plant to the operating waste disposal sites. These allocations are also indicated in Figures 2 and 3. The allocations differ depending on whether a unit or plant is sited or unsited (that is, whether it is in a region that has or does not have a waste disposal site).

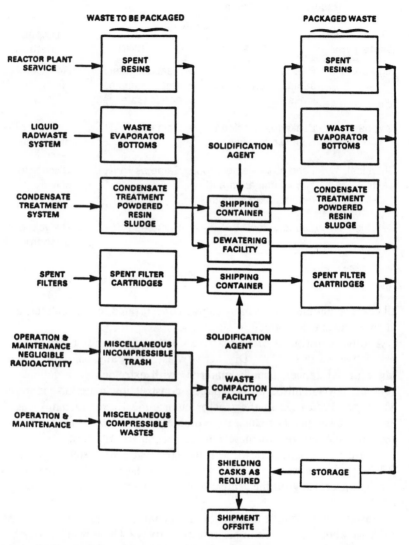

Figure 1. Schematic diagram of a typical radwaste treatment system.

These figures show that twice as much waste (in volume) is produced in a BWR as in a PWR. The data also suggest that the BWRs have had some success in reducing the amount of waste generated over the last seven or eight years, whereas PWRs have shown little change.

The distribution of this waste according to various sources is given in

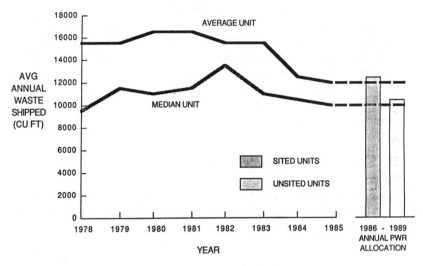

Figure 2.   LLRW volumes shipped from U.S. PWRs.

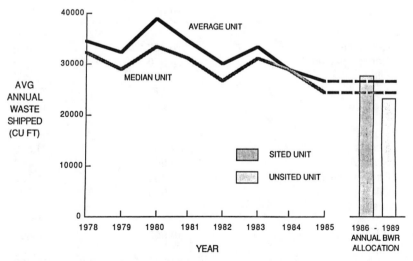

Figure 3.   LLRW volumes shipped from U.S. BWRs.

Table 2 for BWRs and PWRs for 1982–85. In both cases, compacted DAW is the dominant waste source. It is also significant that large amounts of wet waste are produced in BWRs, in contrast to PWRs.

The distribution of radioisotopes in resin wastes for BWRs and PWRs is presented in Table 3. Corrosion product radioisotopes of cobalt are substantial in both cases, with manganese and iron making some contributions in BWRs. In PWRs, the fission products of cesium make significant contributions as well.

RADWASTE MONITORING

Radionuclides may be discharged from a nuclear power plant in gaseous, liquid, or solid form. Each of these must be monitored for radioisotope content in order to assure public health and safety.

Gaseous Wastes

Most prevalent in gaseous radioactivity are radioisotopes of iodine, krypton, and xenon. Gaseous products will separate from liquid at vari-

Table 2.   Waste Summary

Waste Type	Average Cubic Feet/Unit-Year for 1982–85	
	BWRs	PWRs
Dry Waste		
Compacted	10,450	6,350
Noncompacted	8,050	3,700
Filters	100	250
Subtotals	18,600	10,300
Wet Waste		
Resins	2,200	1,500
Sludges	6,000	250
Concentrates	1,750	1,250
Oils	900	300
Misc	50	100
Subtotals	10,900	3,400
Total Average Volume	29,500	13,700

Table 6.3.   Distributions of Key Radioisotopes in Resin Wastes

	BWRs (%)	PWRs (%)
Co-60	50	25
Co-58	4	18
Cs-134	3	10
Cs-137	8	22
Mn-54	15	2
Fe-55	11	5
Cr-51	7	
Zn-65	8	
H-3		7
Ni-63		8

ous locations in the processes within the nuclear power plant. Plants have techniques, such as holdup tanks and charcoal filters, for containing such wastes to permit radioactive decay. They are monitored to ensure that the radionuclide concentrations are within the limits given in 10 CFR 20 and in the guidelines in Appendix I of 10 CFR 50. (CFR is the Code of Federal Regulations.) If concentrations are satisfactorily low, then gaseous wastes are discharged through the stack. If radionuclide concentrations are too high, they must be reduced through natural decay before being released. If regulatory limits are approached, a reduction in plant power output will reduce the production rate of these gaseous radioisotopes.

Liquid and Solid Wastes

Liquid wastes are generally processed by filters, ion exchange resins, and other separation processes to reduce the concentration of radioisotopes in the liquid. Many plants operate under the "near zero release" concept, by which liquids are processed for recycle and reuse within the plant. By filtration or other processes, the radioisotopes contained in the liquid are changed from a liquid to a solid form. The vast majority of radioisotopes released from a nuclear power plant are released as solid wastes. Consequently, the monitoring of these solid wastes for radioisotope content is required principally by 10 CFR 61, which has been amplified by a variety of regulatory guides and branch technical positions issued by the Nuclear Regulatory Commission (NRC).

The NRC, under 10 CFR 61, now requires the division of LLRW into three classes to determine suitability for shallow land burial. These are nominally based on potential radiological hazards, and are determined by the concentration of specific long-lived and short-lived radionuclides. Class A waste contains the lowest concentrations of radionuclides and must meet only minimum waste form requirements. By far the greatest volume of solid waste discharged from a nuclear power plant is Class A. However, the Class A waste also incorporates the smallest amount of radioactivity. Class B and C wastes contain higher concentrations and must meet specified waste form or stability requirements. This generally means that Class B and C wastes must be combined with a solidification agent so that they are encased in a solid matrix. Concrete has long been the most prevalent solidification agent. However, others are now in use, particularly asphalt and thermosetting polymers. Alternatively, Class B and C wastes can be disposed of in high-integrity containers (HICs).

Monitoring

The determination of the appropriate class for any container of waste is established by monitoring the concentration of the radioisotopes and comparing these concentrations to those established in tables in 10 CFR 61. Consequently, the requirements for determination of the various classes through radioisotope monitoring is placed upon the generator of the radioisotopic waste — in this case, a nuclear power plant.

The monitoring of gamma-ray-emitting radioisotopes has been developed and refined through the years. It has become a fairly straightforward process; difficulties arise only when gamma ray energies overlap or interfere with the gamma ray energy of another radioisotope. Analytical methods are available for plant personnel to estimate the total radioactivity in LLRW packages from external radiation field measurements.[3]

It is a bit more difficult to monitor alpha-emitting and beta-emitting radioisotopes that emit no gamma rays. Because the alpha and beta particles emitted from such radioisotopes have a very short range, these radioisotopes must be separated from the gamma nuclides to be monitored. Frequently, radiochemical separations are required to further concentrate and isolate the particular radioisotopes being detected.

Some of the long-lived radioisotopes listed in 10 CFR 61 that determine whether wastes are Class B or C emit only alpha or only beta particles. As noted above, their detection is very difficult and costly for a nuclear power plant operator. Generally, the counting rooms of nuclear power plants are not outfitted to do such radiochemical separations and detections. The determination of the concentration of alpha- and beta-emitting radioisotopes is frequently done by shipping samples offsite to a contractor.

Alternative techniques have been developed recently that use correlations between easy-to-measure radionuclides and difficult-to-measure radionuclides.[1] The latter, of course, are predominantly the alpha-and beta-emitting radionuclides. Wherever such correlations can be developed, they are based on either a chemical similarity between the easy-and difficult-to-measure nuclides or a measured correlation between two such radioisotopes that is consistent in a variety of waste streams and in a variety of plants. Then such correlations can be justified in place of a direct measurement of the difficult-to-measure nuclides.

As the goal of reducing the volume of solid wastes is accomplished, the concentration of radioisotopes in such streams increases. As filter sludges, resins, evaporator bottoms, and other solid wastes increase, monitoring becomes more difficult and complicated. For example, as the concentrations of the radionuclides on ion exchange resins increase, the sample size must be reduced to prevent saturating the electronics in the counting equipment. Also, smaller samples are withdrawn to reduce the radiation dose to the technicians carrying out such a counting process. Many plants will withdraw a very small sample from such high-activity solid waste. However, there is a trade-off in that the homogeneity that must be assumed for the entire waste package in order to use the results from a small sample to monitor the concentration for the entire waste package becomes less and less valid as the sample becomes smaller and smaller.

Techniques have been developed recently to monitor or directly assay an entire waste package rather than monitoring a sample from it. In this case, the monitoring equipment must be taken to the waste package rather than the sample taken to a counting room where the monitors are. Because of the high radioactivity emitted from such a waste package, the monitoring is often done at a significant distance from the waste package. Consequently, computer programs have been developed to account for the distance from the package and also the geometry of the measurement in order to determine the actual radioisotope concentration within the waste package. Such monitors are being refined to improve their accuracy.

Another type of monitor has been developed to determine the concentration of transuranics in a waste package. Transuranics are elements of higher atomic number than uranium and are the source of alpha-emitting radionuclides within a waste package. Some transuranics are also neutron-emitters as a result of spontaneous fissions. Assuming a distribution of transuranic materials in the waste stream, based on the burnup of the fuel in the plant, the detection of the neutrons being emitted from the waste package can be then used to determine the concentration of the various transuranic nuclides present in the waste. Such a system has been developed to monitor the neutrons by surrounding the waste package with neutron detectors and using the count rate, corrected and adjusted for the particular system, to determine the emission rate of the neutrons from the waste package. Figure 4 is a picture of such a system, designed to monitor the neutrons emitted from the transuranics in the waste contained in a 55-gallon drum.

Monitoring characteristics such as those described above can reduce the cost to the nuclear power plant operator, but also increase the confidence and the accuracy of the concentration of the radioisotopes identi-

Figure 4.   Transuranic neutron detection system for LLRW drum.

fied in the waste form. As noted above, these concentrations are critical in determining whether or not the waste form must be stabilized, i.e., solidified or packaged in a high-integrity container, before being shipped for waste disposal at an LLRW disposal site. In addition, the specification of the radioisotope content of any particular waste disposal package is used by the waste disposal operator to determine the inventory of the various radioisotopes in the waste disposal site.

RADWASTE PROCESSING AND VOLUME REDUCTION

Waste Forms

As previously described, there are two forms in which low level radioactive waste is produced in nuclear power plants. The first of these is in the liquid form, termed wet waste, where water-based liquids become contaminated with radioisotopes. The second is in a form called dry active waste (DAW).

DAW, or trash, is the collection of miscellaneous materials that become radioactive through normal plant operations and maintenance. These generally include papers, pieces of wood, tools, plastic floor coverings, and anticontamination clothing such as face masks, booties, shoe covers, and gloves. Some of these materials, such as tools, can be cleaned, recycled, and reused, or can be retained within the radioactive control area as very lightly contaminated material which can be used in that particular state. Attempts are generally made to minimize the amount of material that crosses into the radioactive contamination zone. Once taken in, tools, equipment, and other reusable materials are generally retained within the control zone.

The remainder of DAW usually has extremely low contamination. A large percentage of it will be paper, wood, and plastics. It is usually compacted within a container, such as a 55-gallon drum. Some utilities have gone to a supercompactor that can take an entire drum and compress it down to as little as one-third of its normal size.

In the August 29, 1986, issue of the *Federal Register*, the NRC published a policy statement and implementation plan for handling rulemaking petitions to exempt certain specific waste streams as being below regulatory concern (BRC). This gives radwaste-generators the opportunity to qualify certain waste streams, such as DAW, as having such extremely low concentrations of radioisotopes that they can be considered to be of no threat to the health and safety of the general public. With such exemptions, nuclear power plants could incinerate or dispose of their qualified DAW in a sanitary landfill. This would reduce the cost of waste disposal and conserve

LLRW disposal without increasing the risk to the public. Other waste streams, such as lubricating oils and PWR secondary cooling purification resins could be similarly qualified for BRC.

Waste Processing

Incineration can significantly reduce DAW volume. Currently, only a few organizations are attempting to license incinerators for regional processing sites. These would enable the generators of LLRW to ship their dry active waste to a central location where it could be processed by a number of different techniques, including incineration, to reduce its volume and appropriately package it before it is shipped to a waste disposal site.

As noted in an earlier section, liquid wastes are generated in a number of different locations within a plant. Many plants have the capability of segregating such wastes so as to prevent the mixing of waste streams with significantly different levels of contamination.

In the earlier plants all liquid wastes were processed by evaporators. The evaporator produced purified water that would be recycled within the plant. The still bottoms would retain the impurities in a concentrated fashion, including the radioisotopes. These still bottoms would then be processed as a sludge for disposal.

However, most power plants found the evaporators to be unsatisfactory, due to two features. First, because of the high variability of the chemical nature of the waste streams, such as pH and aggressive contaminants, the heater tubes in such evaporators had a fairly short lifetime, on the order of two years or less. Such frequent replacement requirements led utilities to look for alternatives to evaporators. In addition, it was generally found that these evaporators were somewhat undersized in the design process.

With the general abandonment of evaporators as waste concentrators, nuclear power plants shifted to various forms of filtration and ion exchange. Many systems now use a prefilter to remove particulates from the water, followed by an ion exchange system. Powdered ion exchange resins are used in many plants rather than deep bed resin systems for LLRW processing. Such powdered resins can perform double duty, providing both filtration for extremely small particles and an ion exchange capacity for soluble species in the water.

The purpose of such systems is twofold. First, the water must be sufficiently purified to be recycled within the plant; second, the radioisotopes must be removed from the water so that they can be suitably prepared for disposal.

Recently, techniques have been investigated which would improve the performance of such powdered resin filtration systems. For example, the use of appropriate additives in the waste stream can improve the performance of the powdered resin filtration system by extending its lifetime considerably. This means that much more liquid can be processed through such a filter before it must be replaced. Generally, a filter medium such as this is replaced when the pressure drop across the bed approaches a predetermined value. Consequently, the use of such additives slows the rate of increase of the pressure drop across the bed by preventing the plugging of the very small pores through the creation of larger particulates.

Two of the earliest tests conducted at nuclear power plants are described below. In one, where the average total dissolved solids for the waste stream was 45 ppm, precoat material was modified and a bodyfeed of filter-aid material was added to the waste stream. This resulted in filter performance more than doubling, going from a little over 10,000 gallons filtered per cubic foot of resin material to 22,000 gallons. The latter case, of course, results in more resin sludge material due to the added precoat and the bodyfeed. The resultant improvement in run length for the filter was from 93,000 gallons to 303,000 gallons. At this plant, the cost of disposing of one cubic foot of waste was little more than $200 in 1986. This comes from considering in-plant processing costs (including resin materials, dewatering, and solidification), transportation costs, and burial costs. The changes instituted with this filtration system, and the benefits gained from it in run length, add up to a cost savings of over $300,000 per year at this plant.

A similar test at another plant showed an improvement in filter performance from 1,470 gallons per cubic foot to 11,000 gallons per cubic foot. This resulted in a cost savings estimated to be over $1,000,000 per year for this plant.

Similar techniques which would also reduce the costs and volumes of LLRW are under development. For example, resin materials could be selected based on the chemical characteristics of the particular waste stream, so that the resin was particularly effective in removing the constituents in the waste stream. This is particularly effective in removing radioisotopes of cesium and cobalt, which are the two primary radioisotope constituents in such waste streams.

Advanced Volume Reduction Processes

There are a variety of volume reduction (VR) technologies that have been developed in recent years, often referred to as advanced VR technologies. These include a few that have been mentioned in previous paragraphs, such as the supercompactor and the incinerator, and others, such as the fluidized bed dryer/incinerator, the evaporator crystallizer, the evaporator extruder, mobile incinerators, and mobile thin-film evaporators.[4-7]

The usefulness of these advanced VR technologies to any particular nuclear power plant depends upon a fairly complicated set of circumstances. First, and possibly of primary importance on the list, is the extent of the experience with a particular piece of equipment. Is it at the prototype stage? Design stage? Has it been run? What experience exists with the piece of equipment? Has it been used by the vendor? At a plant? And how dependable are the results that have been predicted for the system?

The second circumstance involves the economics of the process. Such an economic study was completed in 1984, when eight advanced VR systems were considered either from a retrofit or new power plant basis.[8] The economic analysis is developed from a variety of waste streams, including DAW, ion exchange resin, concentrated liquids, filter sludges, and noncombustible trash.

Key factors in the economic analysis include the capital cost of the facility, the cost of money for the particular utility, the cost of transportation for the disposal of the nuclear waste, and the disposal charges themselves. All of these factors can be, and frequently are, distinctive for the particular utility and the power plant location.

Disposal charges have escalated rapidly in recent years, as shown in Figure 5, which demonstrates that the costs for the two major disposal sites in the United States, Barnwell and Hanford, are comparable. (Barnwell is operated by Chem-Nuclear Corporation and is located in South Carolina. Hanford is operated by U.S. Ecology and is located in the state of Washington.) The escalation rate for these disposal charges over the past five years has been a relatively constant 30% per year.

A number of test cases were run, leading to the following conclusions:

- The two most important factors in VR economics are radwaste generation rate and the escalation rate for burial prices; VR is generally more cost effective at BWRs than at PWRs, and at multi-unit stations than at single-unit stations;

- VR can produce significant cost savings in transportation, operation, and storage costs, but these cost reductions are usually the burial savings.

- Variations in construction costs or waste activity produce relatively small changes in economics, but they may be important in borderline economic situations.

- Despite the higher waste activity charges now in effect at waste disposal sites, VR still produces savings.

- Relatively simple trash processors are economical.

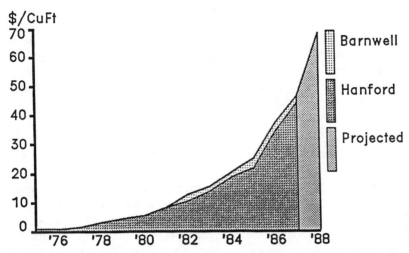

Figure 5.   LLRW burial charges at Hanford and Barnwell disposal sites.

- Evaporator extruders (hot asphalt systems) are economical for plants with moderate or large wet waste generation rates, such as BWRs.
- Fluid bed dryer/incinerators are economical when radwaste generation rates are large enough to overcome their higher purchase and operating costs; however, they have longer payback periods than other technologies.

WASTE DISPOSAL

Siting Plans and Characteristics

States and compacts are presently considering which technologies to select for their LLRW disposal sites. The existing disposal sites all use shallow land burial techniques. A number of alternate technologies have been developed that are being seriously considered and in some cases legislatively required by certain states and compacts.

In general, the performance objectives for LLRW disposal technologies include the following: 1) to separate LLRW from humans; 2) to minimize the availability of radionuclides for dispersal in the environment, 3) maintain the radiation dose to workers at disposal sites at levels "as low as reasonably achievable" (ALARA); and 4) to reduce the impacts resulting from inadvertent intruders at the LLRW disposal site.

Considering these objectives, there are key functional features that are aspects of the technology designs whose presence or absence can directly influence the satisfaction of one or more of the functions required of the disposal technology. There are three distinguishing functional features for LLRW disposal technologies: 1) whether the disposal is to be above or below grade; 2) whether the cover material over the site is to be shallow, meaning on the order of a few meters, or deep, meaning of the order of tens of meters and 3) whether the site has a structure that is separate from the disposal packages as a part of its design, or whether that structure is modular in form.

Technology Selection

Once a design technology satisfies the performance objectives required by the NRC regulations, two other primary factors come into play in the selection of a particular system: economics and politics. In this case, *politics* means the influence brought to bear by politicians or citizens, either for or against certain disposal technologies. These positions may be influenced by the technical performance characteristics of the particular disposal technology, but more often, they are influenced by factors such as the perception of protection or lack of protection that is afforded by certain technologies.

The other factor in the selection process is economics. As a first step, the design basis is determined. The waste is described by its amount, its characteristics, the containers that are used, and the segregation that is required. From this, the facility can be defined in terms of its capacity, size, required physical stability, type of structure, and the intruder protection that will be provided. After this, a particular site can be defined in terms of the buffer zone that is required together with the support buildings and facilities. Specific designs can then be developed with a determination of the total costs associated with each of these technologies.

In general, an economic analysis will require a calendar, which might have the following pattern:

- Site selection — two years
- Environmental impact statement preparation — two years
- Initial construction — two years
- Disposal operation — twenty years
- Site closure — three years
- Long-term maintenance — 100 years

The major costs in this schedule are associated with the disposal operation period of 20 years — the period in which the disposal site accepts the waste from the generators and disposes of it at the particular facility. The resources required to carry out these operations can then be used in a present value analysis to determine the charges that must be allocated for waste disposal.

Currently, there are approximately 12 states or compacts proceeding toward the development of an LLRW site. Of the three sites that are presently open and receiving LLRW for disposal, two (Hanford and Barnwell) receive about 90% of the approximately 2 million cubic feet per year (57,000 cubic meters per year) of LLRW being disposed of from all nonmilitary sources (nuclear power plants, academic and research institutions, hospitals and medical institutions, industrial organizations, and others). The increase in burial costs noted in Figure 5 has resulted in a substantial increase in the use of VR by LLRW generators.

This increase in VR tends to reduce the total volume of increase LLRW produced in any one year. A substantial increase in the number of waste disposal sites would significantly reduce the amount of waste being disposed at any particular site. This will generally produce an unfavorable or more costly economic situation as depicted in Figure 6.

As an example, consider a shallow land disposal facility with a capacity of 240,000 cubic meters. This is a reasonable capacity for ten disposal sites operating for a period of 20 years at the current generation rate. Suppose that the generators shipping to this particular site were successful in reducing their volume by a factor of 2. Since operating costs are the key cost for the LLRW disposal site operator, the effect would be the

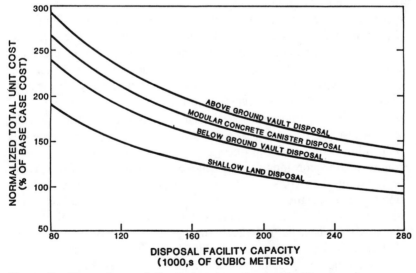

Figure 6.   Dependence of disposal cost on disposal facility capacity.

same as reducing the capacity of the site by a factor of 2. From Figure 6, the operator would be required to increase fees by 50% in order to cover his cost of operation.

For the LLRW generator, the investment is made in VR equipment to reduce the volume of generation by a factor of 2. However, the increase in disposal costs means that rather than saving 50%, the generator would only save about 25% in disposal costs. This, plus the cost of the installation and operation of the new VR equipment may mean an overall increase in costs or a marginal reduction in cost. Nonetheless, most generators must proceed with such VR to protect themselves from cost increases at the waste disposal site that would result from other generators reducing their waste volumes. The result is that the waste generators find themselves on a bit of a treadmill, running at full speed to simply keep pace with LLRW disposal economics.

One solution to this seeming dilemma, is to reduce the number of waste disposal sites from the ten or so presently proposed down to a more reasonable number. With reduced disposal sites, costs would be reflected far to the right and off the scale of Figure 6, where changes in costs are relatively insensitive to the volume of waste received at the disposal site. Currently, there appear to be very few efforts directed toward reducing the number of disposal sites being promulgated in the United States.

Historically, waste disposal charges have been based on the volume that is shipped for disposal. Recently, surcharges have been added, based on the amount of radioactivity that is present in the system. The volumetric charge was based on the sensitivity of the locality toward minimizing the amount of waste that required burial site capacity. The surcharge may reflect an awareness that the potential hazard to the populace was associated not with the volume of the waste, but rather with the amount of radioactivity contained in the waste. Or it may simply reflect an additional opportunity to increase charges.

It is important to consider what benefits might be derived from the various disposal technologies. Such benefits are usually determined by reductions in radiation exposures calculated for individuals from environmental transport models. These models generally consider the precipitation (rainfall) at the site and the various pathways that such precipitation can take into the disposal facility itself. The release of radioisotopes as a result of this infiltration is then determined. Calculations of radioactive doses received by individuals through such models generally show that even the highest dose a person might receive is only a small fraction of the normal background dose that a person receives.

OVERVIEW

It is important to realize that the LLRW process must be considered as an entire system. As illustrated in Figure 7, starting with the source of radioisotopes and proceeding through the processing, packaging, transportation, and disposal of these radioisotopes, most of these steps will also have significant interaction with other steps in the process. Basic to this analysis is identifying the sources within a nuclear power plant. Any reduction in sources will have significant effects on the other processes. The processing of such sources must be related to the sources themselves, their chemical nature, concentration and production rate. But, in addition, the packaging in which the wastes are shipped to a disposal site, whether it be solidification or other processes, will be influenced by the processing that takes place. For example, if concrete is used as solidification media, the processing must produce a form that is solidifiable by concrete. In turn, the package must produce sufficiently low radiation fields that it can be transported to the disposal site with minimal radiation exposures to the waste handlers and transporters. And finally, the disposal site will have certain rules regarding the inventory of radioisotopes present in the package and the waste form itself. Such inventories would be determined by the monitoring procedure.

Consequently, any analysis of the individual components in this system, whether from an economic consideration, a regulatory consideration, an

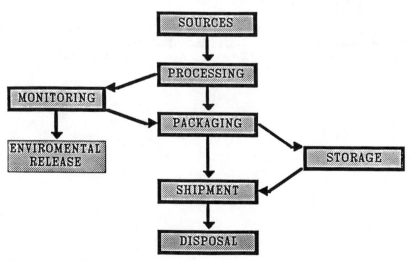

Figure 7. LLRW management system.

operations consideration, or any other base, must certainly consider the full range of the process for appropriate analysis.

Nuclear power plants will undoubtedly continue in the near future to be a major source of LLRW. (An excellent snapshot of the state of affairs for both HLRW and LLRW in early 1987 is found in the Nuclear News.[9]) The nuclear power industry will be looked to frequently for leadership in the various areas that have been discussed in this chapter: source control, waste processing and monitoring, waste packaging, transportation, and waste disposal.

REFERENCES

1 . Electric Power Research Institute. "Radionuclide Correlations in Low-Level Radwaste," EPRI NP-4037 (1985).
2. Electric Power Research Institute. "Identification of Radwaste Sources and Reduction Techniques," Vol. 13, EPRI NP-3370 (1984).
3. Electric Power Research Institute. "Determination of Waste Container Curie Content from Dose Rate Measurements," EPRI NP-3223 (1983).
4. Electric Power Research Institute. "Advanced Low-Level Radwaste Treatment Systems," EPRI NP-1600 (1980).
5. Electric Power Research Institute. "Non-U.S. Advanced Low-Level Radwaste Treatment Systems," EPRI NP-2055 (1981).
6. Electric Power Research Institute. "Radwaste Incinerator Experience," EPRI NP-3250 (1983).
7. Electric Power Research Institute. "Installation and Startup of the Palisades Nuclear Plant Asphalt Volume-Reduction Solidification System," EPRI NP-3933 (1985).
8. Electric Power Research Institute. "Long-Term, Low-Level Radwaste Volume-Reduction Strategies," Vol. 1-5, EPRI NP-3763 (1984).
9. "Waste Management Today," *Nuclear News*, 30 (3): 44–88 (1987).

CHAPTER 7

The Low-Level Radioactive Waste Dilemma: One Institution's Approach

George R. Holeman and Lawrence M. Gibbs

INTRODUCTION

The dilemma of disposal of low-level radioactive waste (LLRW) is an issue faced by radiopharmaceutical companies and biomedical research and health care institutions as well as the nuclear industry itself. This chapter will focus on the programs and methods employed by one such research and educational institution as it strives to minimize the impact on the research community it serves and yet maintain a reasonable, practical, and cost-effective LLRW disposal program. The type of program and the methods used may not be suitable for all applications, and the authors stress that each institution must determine which specific aspects and techniques best meet the goals and objectives of that organization.

LLRW DISPOSAL OPTIONS

The most prevalent disposal option used, and the one which has led to the current problems, is the shipment of materials offsite for burial in a radioactive waste landfill. A review of onsite options is needed at this time.

Straub, in an early comprehensive review of LLRW, introduced the subject by detailing two widely used options for handling nonradioactive industrial wastes, namely "dilute and disperse" and "concentrate and contain."[1] He added a third option for radioactive waste, "delay and decay." Today, most categories of options for disposal of LLRW still fit

Low-Level Radioactive Waste Regulation: Science, Politics, and Fear, Michael E. Burns, Ed., © 1988 Lewis Publishers, Inc., Chelsea, Michigan – Printed in USA.

into this structure. However, all disposal options must be considered with the ultimate goal of maintaining exposures, both occupational and environmental, as low as readily achievable (ALARA). ALARA is a concept utilized by the Nuclear Regulatory Commission to implement its regulations and is usually a fraction of the statutory exposure limits.

Dilute and Disperse

A number of possibilities for LLRW disposal fit into this category. These include the dilution and dispersal of liquids to the environment, the grinding and dilution of biodegradable waste, allowing gaseous materials to escape into the air, and the incineration of volatiles, some solids, and biological wastes, depending upon the isotope and compounds involved. These options are limited or controlled by the level of radioisotope concentration at the release point or the nearest population point. Dispersion is the key element in this process. However, reconcentration or bioaccumulation within the environment after release must always be a consideration with this method.[2]

Concentrate and Contain

Possible options here include compaction, shredding and crushing, filtration, ash from incineration, desiccation of biological materials, evaporation, and solidification of liquids. The concentration mechanism reduces the volume of waste to be handled; however, it also increases the radioactivity concentration levels.

Methods of containment have been reviewed extensively, and shallow land burial of LLRW appears to be the most viable containment option at this time.[3] However, containment of low-level waste need not be considered permanent containment, but may be thought of as a controlled release through dilution and dispersal in the soil.

Delay and Decay

With the interruption of burial site availability, more institutions have turned to onsite storage to allow for decay of short-lived radioisotope waste. Hospitals and medical institutions using radioisotopes with half-lives less than 60 days have been particularly fortunate, because the bulk of the radioactive waste may be held for periods of 10 half-lives or more, thus allowing for essentially complete decay of the radioactivity before disposal as nonradioactive materials.[4,5]

Waste generators with longer-lived isotopes may also use onsite storage by building semipermanent LLRW storage areas.[6] Reactor power plants have moved in this direction, with "life of the plant" storage of LLRW being suggested. Radiopharmaceutical manufacturers provide essential medical care products, and in the process generate significant quantities of low-level waste. Contingency plans of the manufacturers must include provisions for long-term storage of the waste.[7]

Obviously, there is a limit to which decay before disposal can be utilized effectively. Isotopes such as H-3 and C-14, with half-lives of 12.3 years and 5,730 years respectively, are present in the bulk of the institutional waste being generated.[8] Efforts by regulatory agencies have resulted in reclassification of certain concentrations of H-3 and C-14 in specific waste streams so they may be disposed without regard to their radioactive content.[9] However, appropriate disposal must still consider the biological and chemical constituency of the material for appropriate disposal.

ISOTOPE USAGE AT YALE

Yale University, in New Haven, Connecticut, is a leading educational and research institution, with a total enrollment of over 10,000 in the undergraduate, graduate, and professional schools, with teaching and research conducted by 2,200 tenured and nontenured faculty.

The importance of radioisotope usage for research and diagnostic purposes cannot be overstated. As an investigative tool, radioisotopes have, over the past 30 years, significantly advanced the capabilities of researchers and clinicians, especially in biomedical applications.

The Yale University Radiation Safety Committee establishes university radiation safety policies; the Health Physics Division implements the policies and procedures, and provides radiation safety services to the research community. Under the university's Broad Research License issued by the Nuclear Regulatory Commission (NRC), approximately 250 principal investigators at Yale are authorized to use radioactive materials in their laboratory research programs. Before receiving authorization, the university's Radiation Safety Committee reviews the investigator's training and experience with radioactive isotopes, the research facilities, and proposed research protocol. These approved investigators conduct research in approximately 1,000 laboratories on campus. The personal monitoring program of the Health Physics Division covers over 1,500 Yale research personnel.

Research with radiation and radioactive materials at Yale University began around the turn of the century, with the earliest discoveries of X-rays and naturally occurring radioactive materials. Widespread use of radioactive materials as tracers of biochemical compounds, a particularly useful diagnostic and research tool in biomedical research, led to the initial increase in radioactive waste volumes during the 1950s and 1960s (Figure 1). The shipment of radioactive wastes for burial began with the opening of the first commercial radioactive waste burial site in 1963 by Nuclear Engineering Corporation in Kentucky.

The predominant radioactive isotopes used in biomedical research are tritium (H-3) and carbon-14 (C-14). In addition, other continuously popular isotopes include phosphorus-32 (P-32; 14.3-day half-life) and iodine-125 (I-125; 60-day half-life). Besides these most popular research tools, sodium-22 (Na-22; 2.6-yr half-life) and a variety of isotopes in the form of microspheres are used in small quantities. (Microspheres are radiolabeled tracers designed especially for blood circulation studies in animals. They are ion exchange beads such as styrene-divinyl benzene copolymer, uniformly labeled with a radionuclide and coated with a polymeric resin. Microspheres are labeled with a wide selection of radionuclides that are readily detected by gamma-counting techniques.)

Each of these materials presents different disposal problems. H-3 and C-14 cannot benefit from the hold-for-decay option because of their long half-lives. If the H-3 and C-14 wastes are in the form of solids, then burial in a secure waste site is the only option. P-32 waste can often be held for decay unless it is mixed with longer-lived material. The longer, 60-day half-life of I-125 makes the hold-for-decay option less desirable, as holding periods of up to 2 years must be considered. Na-22, a high-energy gamma emitter with a half-life of over 2 years, is most difficult to store, and disposal by approved landfill burial is the most likely option.

Segregation of the various waste streams is extremely important if the different options are to be implemented. The more segregated the waste is at the generation site, the laboratory, the easier it is to implement various disposal options.

LOW-LEVEL RADIOACTIVE WASTE DISPOSAL PROGRAM

Solid Waste

The amount of low-level waste generated at Yale (Figure 2) parallels the growth in both the amount of research conducted and the increased use of isotopes in research during this time period. The types of disposal options used in the past at Yale were similar to those at most research institutions, and included holding for decay and sewer disposal when possible, or shipment offsite for burial. With the rapid expansion of

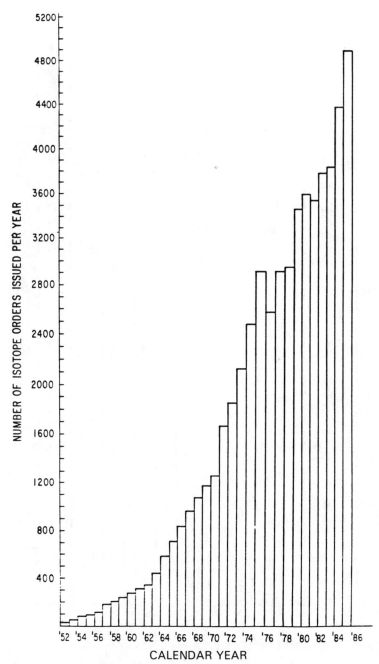

Figure 1. Radioactive isotope requisitions, Yale University.

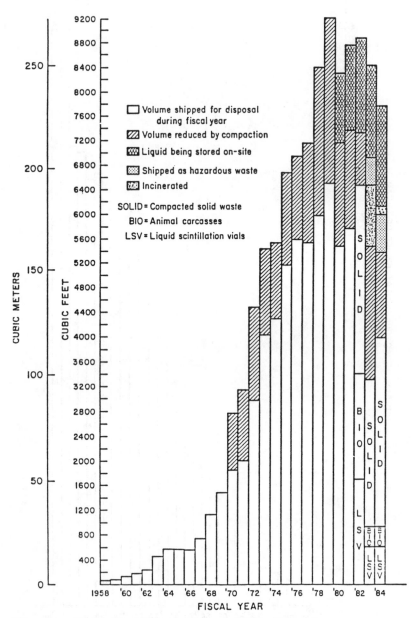

Figure 2. Radioactive waste management, Yale University.

biomedical research programs at Yale during the late 1960s and early 1970s, compaction of dry waste became a very attractive volume reduction consideration. The first compactor was obtained in the early 1970s with a projected capital recovery period of three years. In fact, the compactor expense was recouped in less than a year and a half. The success of this program, coupled with increasing dry waste volumes, led to the purchase of a second compactor in the late 1970s. On the average, compaction has produced a fourfold reduction in the volume of dry waste sent offsite for burial.

Research animal carcasses represent another type of solid waste that may be disposed of through a number of methods: incineration, onsite burial under special NRC and local or state health permit conditions, maceration and subsequent disposal as liquid to the sewer system, preservation and storage for decay, and transfer to a licensed burial site. Carcasses containing C-14 and H-3 may be incinerated, as the C-14 will be retained in the ash as carbon or released as carbon dioxide, and tritium will be released as a gas or as tritiated water. Incineration may also be considered for disposal of carcasses containing short-lived radionuclides in activities up to several millicuries. Those containing isotopes with half-lives of weeks to months may be held for decay for 10 half-lives, or longer when possible, and then declared nonradioactive before incineration. Waste with large quantities of radioactivity may have to be held longer to allow the activity levels to decay to background. Carcasses with long half-lives should be stored in a preserved or frozen state until shipped to a disposal site. Where the carcasses contain large activity levels, microspheres, or long-lived isotopes, consideration should be given to whether incineration, burial, or transfer to a licensed disposal company is preferred. Residual ash may contain radionuclides and should be treated cautiously, as the total volume of ash may be contaminated and may constitute a larger volume than the original carcasses.

Liquid Wastes

A significant portion of low-level radioactive waste from the university's biomedical research facilities is in the form of liquid scintillation counting waste. Under NRC regulations, certain scintillation liquids may be disposed of without regard to their radioactivity,[10] under what is known as the de minimis exemption criteria. In this way, the deregulated waste stream is eliminated from the NRC disposal regulations. However, many of the solvents used in scintillation fluids are flammable, and their disposal is regulated by the Environmental Protection Agency under the hazardous chemical waste regulations.

The radioactive scintillation fluid waste still represents a problem. Although not currently in use at Yale, distillation methods have been developed and used that allow for the separation of the hazardous solvent from the rest of the scintillation cocktail mixture, with the residual radioactivity remaining in the nonsolvent matrix.[11,12] The solvent is separated from the rest of the mixture by spinning band distillation. It was found that with H-3 and C-14, the radioactive isotopes were primarily attached to nonvolatile materials or materials with high boiling temperatures, and remained in the stillpot after distillation. The recovered solvent is free of radioactivity and can be reused as a reagent chemical. Sewer disposal of the remaining materials is, then, one possible option if the activity levels are low enough. If the activity is still above regulated levels, the remaining radioactive matrix has been reduced in volume by a minimum of 30%-40% and may be safely stored, as the hazardous chemical component has been removed. This method shows much promise for both volume reduction of the radioactive material and removal of the flammable hazard for storage consideration.

When the available burial sites first began restricting liquid scintillation vial wastes, the Radiation Safety Committee reviewed methods to minimize the generation of waste liquid-scintillation fluids. Emptying used vials was investigated as one possible alternative. However, the labor-intensiveness of this operation, combined with the increased risk of exposure to the radioactivity and the scintillation solvent, prohibited this method from gaining much acceptance. Instead, the Radiation Safety Committee developed a policy that required all investigators to switch to the smaller (7-mL to 10-mL) minivials from the larger 20 mL vials. Increased sensitivity of scintillation counters as well as new diagnostic techniques have made this switch technically feasible.

REGULATORY ACTIONS AND UNIVERSITY RESPONSE

In 1980, the Low-Level Radioactive Waste Policy Act was passed by Congress. This legislative action placed the responsibility for low-level radioactive waste disposal on state governments and set a time limit of January 1, 1986, for establishment of regional disposal sites. Regions with existing disposal sites at that time could then exclude radioactive waste from states outside their respective regions. In reality, no new sites have been developed to date. There are only the three previously existing sites, one in South Carolina, one in Nevada, and the other in Washington state. Most institutional waste goes to the site in Hanford, Washington. Recent amendments to the 1980 Act extend the availability of these three sites until 1992. However, the amendments allow the sites to place significant restrictions on generators shipping waste to the sites. The facilities are also allowed to place surcharges on waste volumes shipped to the site for wastes generated from outside the region.

In 1985, university-authorized research generated about 8,000 ft^3 of low-level radioactive waste. Anticipating a constant waste volume and the surcharge system established by the recently passed amendments to the Low-Level Radioactive Waste Policy Act of 1980, the university projected burial surcharges of $80,000 for 1986 and 1987, $160,000 for 1988 and 1989, and $320,000 for 1990 and 1991. Since these costs are directly related to the amount of waste materials shipped for burial, there is an obvious incentive for Yale to reduce its waste volume even further. Other conditions of the act, including the possibility of restricted or denied access, prompted the university to review its entire radioactive waste program in early 1985.

Committee Formed

A task force was formed and chaired by the University's deputy provost. A complete review of the existing radioactive waste disposal program was undertaken, as well as discussion of possible new alternatives. Unsure of the outcome of congressional debate regarding the amendments to the 1980 act, Yale proceeded on the assumption that eventually the institution would have to accommodate management of a larger proportion of the waste generated, and for longer periods of time. Table 1 represents a listing of volume reduction factors that were considered when evaluating possible options. The review of possible alternatives and ensuing discussions led to the conclusion that increased volume reduction and preparation for onsite long-term storage best fit the needs and restrictions of the university.

These discussions involved not only the technical feasibilities of various options, but also the potential political ramifications of any particular option. Public concerns about all things nuclear are resulting in increased pressure for more restrictive regulations and increasing concerns about the risks involved in the transport and disposal of radioactive materials. Significant public resistance to the proposed regional sites may delay the opening of such facilities even further.

Increased Volume Reduction and Long-Term Storage

Once the two categories above were identified, investigation of methods to maximize the impact of each were conducted. In volume reduc-

Table 1.   Volume Reduction Evaluation Factors

Design Basis	Operating Factors
Application	Operating History
Process Flow Sheets	Performance and Maintenance Data
Equipment Configuration	Limitations
Pretreatment Requirements	Reliability
Capacity	Availability
Materials Processed	Maintainability
Materials of Construction	Safety
Operating Parameters	Occupational Radiation Exposure
VR Capability	**Cost Factors**
Decontamination Factor	Equipment Capital Cost
Secondary Waste Streams	Installation Cost
Auxiliaries Requirements	Operating Cost
Parameter Descriptions	Processing Cost
Secondary Waste Stream	Cost Effectiveness
Product Characteristic	**Applicability to LLRW**
Operating Conditions	Wastes
Accidents/Upset Conditions	Design Factors
	Retrofit/New Installation
	Compatibility
	Operability
	Licensability

Source: Trigilio.[13]

tion, compaction of dry waste was already being utilized. Liquid scintillation fluid waste, on the other hand, was not being reduced in volume to the extent feasible. The scintillation fluid waste vials were being segregated into radioactive and deregulated streams, with the latter being shipped as hazardous chemical waste. However, the radioactive scintillation vials still constituted approximately 15% of the total volume of waste generated. The proposal to provide long-term storage for quantities of radioactive liquid that is also quite flammable were not attractive. The university is in an urban area and has no easily accessible isolated places in which to locate a radioactive storage area. In addition, the political considerations involved in siting such a facility in a location distant from the campus, but in another municipality, were formidable.

An onsite location was identified, although it would need extensive refurbishment to meet the criteria for storage of flammable and combustible materials. Investigation of methods for the removal of the scintillation fluid from the vials was also undertaken. A variety of separation methods were investigated. These included manual separation by the research investigator in the identified laboratories; manual separation in a centralized location; mechanized vial emptying, either by tipping the vials or by vacuum removal; incineration of vial and contents; and mechanical vial shredding and separation of the liquid from shredded solids.

Pouring by hand, either at the site of generation (in the lab) or after consolidation of the waste, involved the physical emptying of the scintillation vial contents into common containers. It was believed that this would reduce a 55-gal drum of scintillation-counting vial waste by up to 80%. However, the process is very labor-intensive and would require a well-controlled work environment as well as significant employee monitoring for both radiation and chemical exposure.

The possible use of mechanized tray tippers was suggested. This would involve the placement of the scintillation vials back into the "flats" or cardboard cases in which they are bought and delivered. A wire grate could then be placed over the flats, after removal of the caps, and the flats turned upside down to allow the fluid to drain out. Again, this method required extensive labor by both the lab personnel (to replace the vials into the flats) and the personnel emptying the flats. In addition, engineering and exposure monitoring controls would have to be implemented.

In all of the methods investigated, a major concern was the toxicity and flammability of the solvent component of the scintillation cocktail. When we made our study of alternatives, a new scintillation fluid was just being marketed that used a solvent with a high flash point and low toxicity. The solvent's properties were evaluated by the EPA, which concluded that the substance was nonhazardous for chemical waste disposal purposes.

An in-house study was conducted to evaluate the efficacy of the new scintillation cocktail for the types of research being conducted at the university. An evaluative study of the new cocktail was first carried out by the Health Physics Division. This initial study indicated that the cocktail appeared to be a viable substitute. Researchers were then requested to use the new cocktail and evaluate its efficiency and results compared with those of the cocktails they were currently using. The results of these evaluations indicated that the new cocktail worked as well as the others being used. Based upon these in-house experiments, all scintillation cocktail users are required to use the newer cocktail with the nonhazardous solvent component. Those researchers who do not switch will be assessed an additional waste charge and will have to demonstrate a specific research need to use a different scintillation fluid. This may be simply that the cocktail does not work for their intended purpose, or they might demonstrate that the new cocktail is not sensitive enough for the intended research purpose.

The decision to convert to the new type of cocktail was disseminated to the research community via the provost's office, and was well received by the research staff. Individuals have demonstrated a need to use different scintillation fluids and have been accommodated. One specific user of filters has systematically shown efficiencies varying with time and has been given several individual exemptions to the changeover. However, these requests have been very limited in number.

The use of a less hazardous scintillation fluid, combined with the need to further reduce scintillation waste volumes, made the decision to purchase a vial shredder much easier. Personal communication with another research institution that has been successfully using such a vial shredder for over four years was a major contributing factor to the decision. The volume reduction achieved by removing the liquid from the vial is estimated to be at least six to one, although larger volume reductions may be achieved, depending upon the vial size. This program is anticipated to be operational in 1987. To minimize the construction costs associated with such an operation, the vial shredder will be placed outdoors, with overhead protection from the elements. Based upon current projected use requirements, the shredder will operate at only 20% of capacity; however, payback is still anticipated to be within six months of use.

Space is currently being remodeled to accommodate the long-term storage requirements for both a hold-for-decay program and in anticipation of increased restrictions or exclusions by the current burial sites. The storage site will be able to accommodate our institutional needs during development of regional radioactive waste facilities, which may take six to ten years for the Northeast Region. The development of long-term storage facilities, either onsite or near sites of generation, is one option that should be given serious consideration by all generators of low-level radioactive waste. Current radioactive waste site restrictions and surcharges, as well as foreseeable delays in the development of regional low-level waste sites, makes long-term onsite storage a very real alternative. Those who do not plan for such an occurrence may be left with few other options.

SUMMARY AND COMMENT

The discussions and planning conducted by the university task group were a thorough and productive process that will make options available to the university for the safe and efficient management of the low-level radioactive waste being generated at the institution. The decisions made to proceed along the established course were not easy ones. However, to prevent the interruption of the university's major goals, namely education and research, the changes were deemed necessary and given high priority.

Issues and considerations that should be addressed when investigating alternative volume reduction or treatment methods are included as appendices to this chapter. Although the lists are not all-inclusive, they do provide some guidance in this process. In addition, Table 1 can serve as a checklist and aid for those contemplating volume reduction methods.

Political forces on the regional and national levels will determine the availability of commercial disposal sites. Each generator must be in a position to accommodate interrupted access to the existing disposal sites, at least until the goals of the Low-Level Radioactive Waste Policy Act are met and each region has developed its own facilities.

APPENDIX A

Questions to be Considered for Volume Reduction Treatment Methods[13]

Included below are many of the questions that should be considered in attempts to develop volume reduction options. Though not all-inclusive, they do provide some guidance.

A. Nature of the Waste

1. What are the characteristics of the input waste stream?
2. Is the waste a liquid, emulsion, slurry, sludge, wet solid, or a bulk solid?
3. What is the chemical and radioactive composition?
4. Which components are potentially hazardous or toxic?
5. What are the physical properties of the waste stream (viscosity, boiling point, vapor pressure, radioactivity concentrations, flammability, etc.)?
6. What is the volume or mass of waste that will require treatment?
7. Where does the waste stem originate and at what points can it be intercepted for treatment?

B. Objectives of Treatment

1. What is the present treatment/disposal method and why is this unacceptable?
2. What are the objectives or goals of treatment in order of priority (purification, detoxification, volume reduction, regulatory compliance, concentration, etc.)?
3. What air, water, and other environmental quality regulations must be complied with?
4. What are the chemical and physical property requirements for recycling, reuse, or disposal of the output stream?

C. Technical Adequacy of Volume Reduction Process Alternatives

1. Which volume reduction processes, alone or in combination, can meet the treatment objectives?
2. If a sequence of volume reduction processes is needed, are the processes technically compatible?
3. Can the major volume reduction process selected for meeting the objectives handle the waste stream in its existing form? If not, can the waste stream be altered to a form amenable to the treatment process?
4. Which key processes are technically attractive and achievable?
5. Can an existing process be modified to meet the treatment needs?

D. Economic Considerations

1. What are the capital investment requirements for process implementation?
2. What are the operational costs?
3. What is the cost impact of necessary environmental controls?
4. How do projected alternative treatment costs compare withthe costs of current disposal methods, and with each other?
5. What is the payback period for the proposed treatment technology?

E. Overall Evaluation

1. What factors are most important in selecting a treatment process?
2. Are there standardized, state-of-the-art processes that can be used for baseline comparisons?
3. Are projected costs "reasonable"?
4. Does any process have clearly desirable/undesirable features that make it attractive/unattractive for treating a particular waste stream?

APPENDIX B

Volume Reduction System Design Specification Considerations[13]

Once a volume reduction process has been selected, or is even under consideration, the system's design characteristics and needs should be delineated and evaluated. These system considerations include:

A. Scope: identification of the general requirements of the system and the scope of services requested.

B. Site and facility conditions explicitly delineated.

C. Applicable standards cited: equipment and/or services comply with all pertinent regulations regarding the place of installation, with specified codes and standards referenced.

D. Functional system description: specification of general system requirements and interfaces with other plant systems.

E. Equipment and services furnished by the vendor: provides a major equipment list while leaving final equipment list and system configuration to be developed by the vendor. For a volume reduction system, a request for full technical support, including licensing, installation, and startup, would typically be requested.

F. Design and operating conditions: identification of equipment location, fluid conditions, required effluent conditions, product conditions, and any unique factors that might influence a volume reduction system standard design.

G. Design requirements: identification of the volume reduction system as being designed to standards and other criteria. Requirements for remote operations, decontamination, flushing, special radiation exposure design features, and specific component design features might be required.

H. Construction and fabrication requirements: specify quality standards for equipment tabulation, welding, materials, handling features, structural requirements, and other factors deemed important.

I. Equipment arrangement: addresses the actual requirements for placing a volume reduction system in an available building, special radiation exposure design features, and such items as personnel access and equipment removal.

J. Controls and instrumentation: identifies requirements for remote operation, local control, and instrumentation location and performance.

K. Quality assurance: an appropriate inspection/quality control system will be required to comply with the construction and fabrication specifications.

L. Inspection and tests: required to conform to applicable codes and standards as a minimum. Final acceptance testing is typically performed in the field after the vendor's equipment has been delivered and integrated with purchaser-supplied equipment.

M. Drawings and documentation: requirements for equipment outline drawings, assembly and detail drawings, and complete documentation for operations and maintenance.

N. Information required with the bid: includes operating data, equipment parameters, and special analysis.

REFERENCES

1. Straub, C.P., "Low Level Radioactive Waste," United States Atomic Energy Commission, Division of Technical Information, 1964.
2. National Committee on Radiation Protection, "Radioactive Waste Disposal in the Ocean," NBS Handbook 58 (1954).
3. Holcomb, W.F., and Goldberg, S.M., "Available Methods of Solidification for Low-Level Radioactive Wastes in the United States," United State Environmental Protection Agency Technical Note ORP/TAD764, 1976.
4. Miller, V.L., United States Nuclear Regulatory Commission, letter to all medical licensees, June 4, 1981 (unpublished).
5. Gregory, W.D., "Radioactive Decay as a Waste Disposal Method," paper presented at Health Physics Society Meeting, Las Vegas, NV, June 28, 1982.
6. U.S. Nuclear Regulatory Commission, "Safety Considerations for Temporary Onsite Storage of Low-Level Radioactive Waste" (draft document; no date).
7. Brantley, J.C., "Where and What Are the Wastes?", *Disposal of Low-Level Radioactive Biomedical Wastes*, proceedings of a National Academy of Sciences Forum, November 24, 1980.
8. NUS Corporation, "Preliminary State by State Assessment of Low Level Radioactive Wastes Shipped to Commercial Burial Grounds" (NUS 3440) and "The 1979 State by State Assessment of Low Level Radioactive Wastes Shipped to Commercial Burial Grounds" (NUS 3440 rev 1), both 1980.
9. Miller, V.L., U.S. Nuclear Regulatory Commission, letter to all medical, academic, and industrial research licensees, October 28, 1981.

10. Title 10 Code of Federal Regulations, Part 20, 306.
11. Mangravite, J.A., Gallis, D., and Foery, R., "The Recovery of Organic Solvents from Liquid Scintillation Waste," *Am. Lab.* 15(7):24–34 (1983).
12. Gibbs, L.M., "Recovery of Waste Organic Solvents in a Health Care Institution," *Am. Clinical Products Rev.*, November/December 1983.
13. Trigilio, F., "Volume Reduction Techniques in Low-Level Radioactive Waste Management", USNRC Report NUREG/CR–2206, 1981.

CHAPTER 8

Low-Level Radiation: Cancer Risk Assessment and Government Regulation to Protect Public Health

Don G. Scroggin

INTRODUCTION AND SUMMARY

The use of health risk assessments as the basis for governmental regulation of hazardous substances, including toxic chemicals and radiation, sparks both legal and public concerns. Industry, environmental activist groups, government regulators, and the general public all grapple with the question of how environmental risk to public health is to be determined and what level of risk justifies governmental regulatory intervention.

This chapter first summarizes the regulatory structure developed by the Nuclear Regulatory Commission (NRC) to promote safe storage of low-level radioactive waste (LLRW), pursuant to the Low-Level Radioactive Waste Policy Act of 1980 (LLRWPA).[1] Then follows a discussion of why and how the NRC is likely to rely upon generic guidance from the Environmental Protection Agency (EPA) in utilizing risk assessments to establish standards for determining when certain streams of LLRW may be exempted from regulation because they are "below regulatory concern."[2]

Generically, risk assessments are the scientific tools used by risk managers in making the critical threshold determination required before virtually any regulatory program can be undertaken: does the risk to public health justify governmental action? Because risk assessments contain numerous assumptions and uncertainties, the manner in which the final risk estimates are characterized when transmitted to the risk managers may well determine how these numbers will be perceived. However, vir-

Low-Level Radioactive Waste Regulation: Science, Politics, and Fear, Michael E. Burns, Ed., © 1988 Lewis Publishers, Inc., Chelsea, Michigan – Printed in USA.

tually all such cancer risk calculations are presented as upper bound estimates, rather than calculations of the actual risk to the public. Thus, different agencies and even different scientists within the same agency may use different assumptions and different levels of sophistication in disclosing these uncertainties to the decisionmakers. As a consequence, the assumed risks of exposure to the same pollutant may vary among agencies or among different sections of the same agency. For example, former EPA Administrator Ruckelshaus discovered that within EPA itself, risk assessments by different EPA offices for the same chemical pollutant might differ by three orders of magnitude (up to 1000).

In response to this inconsistency, Ruckelshaus initiated an effort to make more uniform the manner in which risk assessments are used in EPA's environmental decisionmaking. The focus of this effort was the development by EPA of generic risk assessment guidelines, which would establish general criteria for the use of health risk assessments in government regulations.[3] These generic risk assessment guidelines are intended for use throughout EPA and would be expected to be influential with all federal and state agencies with similar mandates to protect public health from risks associated with exposure to hazardous substances, including radiation.

In order to assure that EPA's new generic risk assessment guidelines were scientifically sound, Administrator Ruckelshaus requested that EPA's prestigious and independent Science Advisory Board (SAB) review them and offer recommendations. In response to the Administrator's request, the SAB drafted detailed recommendations, focussing primarily on how the cancer risk assessment guidelines treated the critical issues of explicit disclosure of uncertainties, the range of risk, and the assumptions underlying such risk estimates. The explicit and strongly worded recommendations of the SAB sparked a heated debate within EPA and among other expert scientific agencies within the federal government. This controversy delayed issuance of the guidelines for many months and resulted in EPA Administrator Thomas's addition of significant new language to the final guidelines. The second portion of this chapter discusses that controversy and its implications for future use of risk assessments by government agencies.

Finally, this chapter discusses the radionuclides Clean Air Act test case,[4] currently before the D.C. Court of Appeals. This radionuclides litigation provides a useful illustration of the practical implications of all of these points and is the first case in which the courts will address the difficult question of what level of public health risk is sufficient to trigger governmental action. The precedent of the radionuclides case is expected to influence the government's approach in protecting public health not only from radiation, but from all hazardous substances.

CURRENT REGULATORY OVERSIGHT OF LLRW

The protection of the public health from the potential risks associated with disposal of LLRW was the province of the federal government alone until 1980, when a critical shortage of suitable disposal sites prompted the passage of the LLRWPA, which shifted responsibility for such disposal largely to the states.[5] However, the failure of any states to open new disposal sites after passage of the LLRWPA prompted those three states[6] with operating disposal sites to declare their unwillingness to continue to receive LLRW from other states. This refusal precipitated a potential crisis, since the lack of suitable disposal sites would have shut down several industries that generate nuclear waste by the statutory deadline of January 1, 1986.[7] For example, virtually all pharmacological research, which routinely uses radioisotopes, would have been halted.

In response to this crisis, Congress passed the 1986 amendments to the LLRWPA.[8] These amendments approved seven regional compacts, encompassing 37 states. Until these compacts are effective, the three existing disposal sites will continue to accept LLRW, subject to certain financial incentives for the development of other disposal sites.[9]

The 1986 amendments require the regulation of all LLRW, but provide that the NRC "shall establish standards and procedures" for acting upon petitions

> to exempt specific radioactive waste streams from regulation by the [NRC] due to the presence of radionuclides in such waste streams in sufficiently low concentrations or quantities as to be below regulatory concern.[10]

The amendments state that in establishing such standards and procedures "to exempt the disposal of such radioactive waste from regulation" the NRC must "protect the public health and safety."[11]

In addition, mixed waste — waste that is both hazardous and slightly radioactive — would fall under the jurisdiction of the Resource Conservation and Recovery Act (RCRA), administered by EPA, as well as the authority of the NRC under the LLRWPA and the 1986 amendments. While the House version of the 1986 amendments provided for the development of joint regulations by NRC and EPA to address mixed wastes and avoid the potential for dual (and possibly inconsistent) regulations, consideration of that provision was deferred at the last minute in order to assure passage of the amendments before the deadline of January 1, 1986.[12] Thus, it appears that EPA eventually will be drawn into the determination of whether the contents of mixed waste streams are "below regulatory concern."

The 1986 amendments to the LLRWPA are not alone in requiring an assessment of the risk to public health. Indeed, implementation of virtually all of the environmental laws regulating toxic or cancer-causing pollutants requires some assessment of risk to human health as a predicate for regulation.[13] For example, Congress and the courts have determined that the Clean Air Act requires that before regulating a hazardous air pollutant, EPA must make a determination that a significant risk is posed by levels of pollution actually occurring in the ambient air.[14] As the Supreme Court has held, an agency may not substitute a "theory" for a formal determination that the actual levels of a pollutant in the ambient air pose a significant health risk.[15]

EPA proposed its Risk Assessment Guidelines "for use within the policy and procedural framework provided by the various statutes that EPA administers. . . ."[16] Other environmental statutes, such as the Resource Conservation and Recovery Act,[17] the Comprehensive Environmental Response, Compensation and Liability Act (CERCLA, or Superfund),[18] the Clean Water Act,[19] the Federal Insecticide, Fungicide and Rodenticide Act (FIFRA),[20] the Toxic Substances Control Act (TSCA),[21] and the Safe Drinking Water Act (SDWA)[22] have similar requirements that an assessment of risk precede regulation.

EPA'S GENERIC RISK ASSESSMENT GUIDELINES

In determining whether a waste is "below regulatory concern," the NRC will face issues similar to those addressed for the regulation of any carcinogenic substance—namely, what level of exposure to the general public is significant enough to warrant government regulation? In addressing this question, an agency must rely on scientific risk assessments designed to calculate the risk to public health associated with exposure to LLRW. The resolution of the current controversy surrounding EPA's generic risk assessment guidelines will determine how the agency approaches this question.[23] EPA developed these risk assessment guidelines in order to set general criteria for the adequacy of the risk assessments that the agency uses as a basis for its environmental regulatory programs.

EPA's proposed generic risk assessment guidelines proved quite controversial, and final issuance was delayed for almost a year because of continuing debate—both within EPA and between EPA and other expert agencies—over the proper scientific approach to cancer risk assessment. A major component of this debate became clearly focused when EPA Administrator Ruckelshaus requested EPA's prestigious and independent Science Advisory Board to formally review the proposed guidelines.

In response to the Administrator's request, the SAB undertook an intensive review, and drafted specific and strongly worded recommendations. Of particular concern to the SAB was the failure of the draft proposal to emphasize the need for explicit disclosure of the underlying assumptions and the uncertainties in all risk assessments. To emphasize the importance of this recommendation, the SAB drafted specific language for incorporation into the guidelines, disclosing explicitly that the risk assessment numbers presented are in fact upper bounds and are not intended to represent the real risk, which most probably lies somewhere between the upper bound and a lower bound (which is expected most often to be zero). This requirement of full disclosure of uncertainties is well recognized by scientists as fundamental to the scientific method, yet it will trigger a basic change in the manner in which EPA currently conducts risk assessments as the basis for regulation.

EPA's generic risk assessment guidelines represent a final stage in the agency's formal adoption of the risk assessment/risk management decisionmaking structure advocated by the National Academy of Sciences[24] as the framework for implementing environmental statutes premised on an assessment of risk. EPA's adoption of this methodology[25] was in response to EPA Administrator William D. Ruckelshaus's mandate that in agency rulemakings to protect human health, "our science analysis be rigorous and the quality of our data be high."[26] In this framework, risk assessment is the quantitative estimation of the magnitude of health risks and the uncertainties that accompany them; risk management is the policy judgment of what is an acceptable risk to society. The far-reaching implications of EPA's adopting this framework in agency rulemakings have become apparent with the SAB's recommendations.

Requiring explicit disclosure of the uncertainties associated with risk assessments could substantively change threshold agency decisions regarding which hazardous pollutants deserve priority treatment and, conversely, which pollutants fail to warrant any regulation at all because the health risk they pose is not significant. Moreover, regulatory actions made without such full and explicit disclosure of the uncertainty information to the decisionmaker and to the public may well be subject to legal attack as arbitrary and capricious. This new perspective on risk uncertainty also presents a valuable tool to industries and agencies as they comply with federal and state "right to know" laws requiring the communication of health risk to workers and the public. And in toxic tort claims, a new awareness of uncertainties may pose additional difficulties to plaintiffs seeking compensation for disease or injuries allegedly caused by exposure to hazardous substances.

APPROACH TO RISK ASSESSMENT

Currently, virtually all health risk assessments present only the upper-limit extreme of an estimated range of risk, and fail to disclose that the actual risk could lie anywhere between this upper limit and zero. The SAB advised EPA that every risk assessment must contain specific language stating the uncertainties and the range in the estimates of health risk, including both the upper and lower limits, rather than just the upper limits currently used. For example, rather than stating that the risk posed by industrial emissions of a certain cancer-causing substance is "10 cancer deaths per year," the risk assessment would now read explicitly that "the real risk is unknown, but is estimated to lie between an upper limit of 10 cancer deaths per year and a lower limit of zero." Whenever possible, the presentation of "most likely or best estimates of risk for use in risk management" also would be required.

The SAB's recommendations represent the final and necessary step in completing the separation of risk assessment from risk management—a separation that the National Academy of Sciences and Administrator Ruckelshaus viewed as essential to the effective implementation of regulatory programs premised on a scientific assessment of risk.[27] Perhaps the most fundamental of the SAB's recommended changes in this regard was the explicit incorporation into every risk assessment of specific language disclosing to the risk manager and to the public the uncertainties associated with the risk estimates. That language points out that the theoretical model used to project cancer risks "leads to a plausible upper limit to the risk," and emphasizes that

> such an estimate does not necessarily give a realistic prediction of the risk. The true value of the risk is uncertain, and for many substances the lower bound estimate of risk is zero. The Agency's procedures lead to a range of risk, defined by the upper-limit estimate from the linearized multi-stage model and a lower limit, which should be explicitly stated. . . . [O]n a case-by-case basis where the data and procedures are available, the Agency will strive to provide most likely or best estimates of risk for use in risk management. This will be most feasible when human data are available and when exposures are in the dose range of the data.[28]

While fundamental to scientific adequacy, this language presents to the risk manager and the public a remarkably new perspective—namely, that the real risk is uncertain but is estimated to lie between two equally likely extremes—zero and the upper bound. For a policymaker contemplating addressing a public health risk with a regulatory program that could impose millions, even billions, of dollars of additional costs on industry and society, this new perspective would seem to be highly pertinent.

Perhaps because of the broad implications of the SAB's recommended language,[29] EPA debated for several months how to address the SAB's concerns before sending, in late 1985, the final Guidelines to the Office of Management and Budget (OMB) for approval, as required by Executive Order 12291.[30] In spite of the SAB's strong admonitions, however, EPA, without explanation, sent the final Guidelines to OMB without the critical language recommended by the SAB.

When OMB discovered that the version of the Guidelines sent for approval failed to address the SAB's concerns, OMB's role as the watchman over economic impact of government regulations took an ironic turn. Recognizing the far-reaching economic impact the SAB's recommendation would have in providing a sound economic and scientific basis for declining to regulate trivial risks, OMB insisted the EPA reconsider including the scientific principles voiced by the SAB. In the intense debate that followed, OMB expressed its "principal concern . . . that in a few critical instances the guidelines do not ensure that the regulatory decisionmaker will be presented with sufficient information to understand the extent of the uncertainty underlying particular estimates of risk."[31]

OMB's concerns were reinforced by comments from the White House Office of Science and Technology Policy (OSTP) and the National Science Foundation (NSF), both of whom expressed grave concerns regarding the potential for misinterpretation of risk assessment information by decisionmakers and by the public, unless the uncertainties were made explicit. Both OSTP and NSF urged that the scientific integrity of the risk assessment process not be compromised by a failure to disclose explicitly the uncertainties and assumptions associated with the risk estimates. OSTP, noting that " [t]he risk may be zero" for environmental exposure to low levels of cancer-causing substances, commented that "quantification of the various sources of uncertainty can be as important as the estimate itself."[32] It would be scientifically "misleading and spurious," OSTP emphasized, to present upper limit estimates of risks as though they represented the real risk, because "it leads to a burial of the uncertainty in the estimates."[33] Similarly, NSF urged that "the risk assessment portion of such risk analysis contain a full disclosure of the range of data so that the relative uncertainty could be assessed."[34]

The controversy was not resolved until EPA Administrator Lee Thomas agreed to add three prefatory paragraphs to the Guidelines, in which he directly addressed the SAB's (and OMB's) concerns. After referencing the SAB's review of the Guidelines, the revised Guidelines now read:

In particular, the guidelines emphasize that risk assessments will be conducted on a case-by-case basis, giving full consideration to all relevant scientific information . . . . [The] Agency scientists will identify the strengths and weaknesses of each assessment by describing uncertainties, assumptions, and limitations, as well as the scientific basis and rationale for each assessment.[35]

Thus, while the SAB's interpretative language does not appear explicitly in the new Guidelines, given the history of the SAB's involvement in their development (and the SAB's congressionally-mandated reviewing role), the SAB's analysis and recommendations offer specific guidance to the courts and to the public for determining the precise requirements of the final Risk Assessment Guidelines.

Perhaps the most striking implication of the Guidelines' requiring explicit disclosure of the uncertainties associated with risk assessments is the potential of that disclosure to influence threshold agency decisions regarding which hazardous pollutants deserve priority treatment and, conversely, which pollutants fail to warrant any regulation at all because the health risk they pose is not significant. Regulatory decisions made with such full disclosure of the uncertainty information should make such actions more acceptable to the public and better able to withstand legal challenge in the courts. Conversely, decisions based on a selective presentation of risk information, contrary to the requirements of the Guidelines, would appear to have a legal Achilles' heel.

How Much Risk?

EPA's regulatory process is complicated by the great difficulty in conducting most health risk assessments. In most cases no identifiable harm to human health has ever been observed from environmental exposure to the hazardous substance in question. In those rare cases where human health effects have been detected—such as in occupational exposures to radiation, vinyl chloride, or benzene—the illnesses or cancer were caused by exposures many times higher than those being considered for regulation.

As a consequence, agencies must rely on risk assessment estimates that are based on experiments conducted with animals at doses sometimes thousands of times higher than those actually encountered in the environment. Assumptions are then made regarding the similarity between the test animals and humans. The health effects at the levels of exposure humans actually encounter in the environment are then estimated using the linear nonthreshold theory, which assumes that any dose, no matter how small, will cause a harm proportionate to that at high levels.

There are significant scientific problems with using only the linear nonthreshold theory, because it tends to overestimate the calculated risks. Yet, advocates of the linear nonthreshold theory argue that prudence dictates use of its "conservative" risk estimates. Other scientists point out that, since there is no evidence confirming or refuting the linear nonthreshold theory at the extremely low levels considered for regulation, other dose-response curves should also be considered.[36]

Some activist groups strongly urge agencies to rely upon risk assessments that use the linear nonthreshold theory and not to disclose the uncertainties and range associated with the estimates of risk in promulgating regulations. These groups advocate regulation of all human exposures to hazardous industrial pollutants, no matter how low the level of exposure actually occurring. For example, in recent comments submitted to EPA regarding the agency's proposed withdrawal of benzene regulations[37] because the health risks were determined to be insignificant, one such organization argued that public health agencies such as EPA should respond to even one "statistical" death with the same urgency with which a police department would respond to "a tip that a dangerous person has threatened to shoot randomly into a Times Square crowd until he kills one person. . . ."[38]

The Decisionmaker's Role

While full disclosure of uncertainties associated with risk assessments is a fundamental requirement for scientific adequacy,[39] risk assessments intended for use in governmental regulations have rarely included this information. The recognition that a regulation, perhaps costing society hundreds of millions of dollars, was prompted by a calculated health risk that is as likely to be zero as it is to be the upper-limit number and that hinges upon unproven theoretical assumptions, may be disquieting to policymakers and risk assessors, as well as to the parties directly affected. As a consequence, risk numbers that are in fact only the extreme upper limit of the estimated range of risk tend to assume a false appearance of precision and certainty, taking on a life of their own in the press and in the minds of the public. More important, however, they create a faulty predicate for rule making.

The failure of current risk assessments to lay out openly and candidly the full range of scientific information available, including the uncertainty information, is precisely what renders the final regulatory actions based upon such risk assessments legally vulnerable. While courts are reluctant to substitute their judgment as to how various factors should be weighed in reaching an agency's regulatory decisions, they are nevertheless quick to condemn as arbitrary and capricious agency decisions based on less than full consideration of all the relevant information. Here, an analogy to the National Environmental Policy Act is particularly apt. Pursuant to that Act, an environmental impact statement (EIS), much like a risk assessment, is required to lay out the range of information and environmental impacts, but not to reach a final decision as to what the policymaker's final action should be.

Similarly, a risk assessment should lay out the range of scientific information, including the range of risk, the uncertainties associated with that calculation, and the explicit information that the real risk is unknown but most likely lies between the lower and upper bounds that are explicitly stated. A decisionmaker might then choose not to regulate a substance for which the uncertainties were large and the data were sparse. Moreover, much as a failure of an EIS to set forth the full range of environmental impacts might render arbitrary and capricious an agency's decision, a failure to consider the full range of scientific information, including uncertainties, in a risk assessment may similarly render an agency's action susceptible to being judged arbitrary and capricious.[40]

Giving the risk manager only the extreme upper bound of the risk is much like giving a manager who manufactures blue jeans only the largest size worn by anyone in the country—having limited resources, he needs more information on the range of sizes and the uncertainty of the data if he is to make intelligent decisions regarding what to manufacture. Decisions based on such incomplete data are not likely to withstand the test of the marketplace; similarly, regulations premised on comparably select and incomplete data may be vulnerable to challenge.

Agency Priorities

Consideration of the uncertainties and range in risk assessments also should be an important aid to EPA's ability to focus its finite resources on the more significant and certain environmental hazards society faces. Otherwise, the agency may expend its efforts chasing risks that are relatively small or based on dubious assumptions, at the expense of addressing more significant risks, whose costs to human health are well established and large. Two hazards might have the identical upper limits of risk, for example, but have quite different scientific certainties associated with how those risks were calculated and how effectively emission controls will reduce the risk.[41]

Thus, exposure to two hazardous substances might each have an upper-bound risk of 10 deaths per year in the United States, but one risk might be well documented with human data, while the other might be based solely on a single animal study at extremely high doses. Moreover, one risk might be able to be reduced to half its initial size with an expenditure of a million dollars, while the other might remain virtually unchanged by such an expenditure. Without the important scientific information regarding assumptions, uncertainties, and range of risk—all of which the SAB has advised EPA to include as a fundamental component of the risk assessment process—EPA as risk manager is hampered in the performance of its most fundamental duties.

RADIONUCLIDES TEST CASE

The hotly contested litigation concerning EPA's regulations to limit emissions of airborne radioactive particles (radionuclides) pursuant to Section 112 of the Clean Air Act pointedly illustrates the value of a risk assessment that fully discloses uncertainties to the risk manager.[42] Former Deputy EPA Administrator Alvin L. Alm recently asserted that industrial emissions of radionuclides pose a health risk that is "minuscule" when compared with other exposures to radiation.[43] A Harvard law professor, Richard B. Stewart, has also observed that in light of the "quite trivial" risks posed by radionuclides emissions, "[t]he entire exercise represents a serious misallocation of administrative resources."[44] On the other side, however, several activist groups have voiced outrage at the agency's consideration of the significance of risk. In the words of one group, "The decision not to set standards that would reduce public exposure to known causes of cancer is a decision to abandon the war on cancer."[45]

A special committee of the SAB has reviewed the risk assessment for radionuclides and concluded that it *"is not an adequate or balanced assessment of the scientific data pertaining to airborne radionuclides, and it cannot be judged as a scientifically adequate basis for regulatory decisions for this pollutant"* (emphasis in original).[46] The SAB further recommended that, instead of presenting only upper-limit risk numbers, the scientific uncertainties "should be expressed as central estimates with lower and upper bounds. . . ."[47] If EPA initially had been able to consider such uncertainty information in the risk assessment, the agency might have chosen to focus its limited resources on more significant risks, or at least would have been better positioned to defend the actions it did take.

Right-to-Know Laws

Federal and state laws also require the communication of health risks to workers and to nearby communities exposed to hazardous substances. OSHA requires employers in the manufacturing sector to inform their workers of the nature of health risk presented by hazardous substances to which they are exposed. Several states have recently enacted broader laws requiring communication of such risks to the communities as well and covering the nonmanufacturing sectors.[48] And recently proposed national legislation would require comprehensive notification requirements for communities near potential chemical hazards.[49]

The new requirements of full disclosure of uncertainties in risk assessments may influence how these risks are perceived by workers and the public. The chief question asked by the public is "Is it safe?" Most frequently, however, science simply cannot answer this question with certainty. Communication of risk to the public, according to former Administrator Ruckelshaus, should convey what the experts really think, and not give a false impression of certainty that does not actually exist as to either the danger or the safety of a particular substance.[50] This is precisely the sort of information the SAB has determined to be fundamental to risk assessment.

Toxic Torts

The new perspective offered by the risk assessment guidelines, with the Science Advisory Board's recommendations, also could have a noticeable effect in the area of toxic torts litigation. For example, in addition to traditional tort causes of action in which plaintiffs claim damages for cancer and other illnesses allegedly caused by exposure to hazardous substances, recently proposed legislation would create a federal cause of action for such plaintiffs.[51]

Courts are now called upon to assess such questions as whether the plaintiff who has cancer is one of the six excess cancer cases predicted by a risk assessment to occur in the United States. Such deceptively precise upper limits of current risk assessments are frequently the only numbers considered. If, rather, a risk assessment stated explicitly that the actual risk is 0 to 6 with a most probable value of 2 and that the real risk is uncertain, a court may view quite differently the likelihood that the plaintiff deserves compensation from the defendant. In particular, the information that the risk is just as likely to be zero as it is to be the upper-limit number of deaths may alter a court's perspective in evaluating the cause of the plaintiff's harm.[52]

For example, the recent settlement in the Agent Orange case, involving alleged health effects resulting from exposure of Vietnam veterans to dioxin, may have been influenced in part by the uncertainties in the risk assessments linking exposure to illnesses and birth defects. On this point, the *New York Times* recently editorialized that the veterans' attorneys "shaped their suit on a shadow that grows fainter in the light of each new health study."[53] And a New Jersey appellate court recently overruled a lower court's damage award, which was based on plaintiff's alleged increased risk of cancer. Specifically, the court held that the evidence did not prove that the "defendant has so significantly increased the 'reasonable probability' . . . that any of the plaintiffs will develop cancer so as to justify imposing upon defendant the financial burden" demanded by plaintiffs.[54]

Overall Impact

In sum, EPA's Risk Assessment Guidelines will provide definitive guidance to EPA and will be an influential precedent for other agencies engaged in regulations to protect human health, such as OSHA, CPSC, FDA, and HHS. The formal recommendations of the Science Advisory Board regarding full disclosure of the uncertainties in agency rule making are fundamental principles of the scientific method and constitute the final and necessary step in EPA's adoption of the National Academy of Sciences' risk assessment/risk management decisionmaking framework. While the SAB's recommendations introduce a new perspective to rule making, coping with uncertainty is crucial to fulfilling the agency's statutory mandates to protect human health from environmental pollutants and, ultimately, to the legal sustainability of its rules.

REFERENCES

1. P.L. 96–573.
2. P.L. 99–240, Section 10.
3. 49 Fed. Reg. 46294–46331 (November 23, 1984), 50 Fed. Reg. 1170 (January 9, 1985), and 51 Fed. Reg. 33, 992 (Sept. 24, 1986). See Scroggin, D.G., "New Guidelines Face Complicated Problem of Risk Evaluation," *Legal Times*, February 23, 1987, p. 20.
4. *Environmental Defense Fund et al.* v. *Ruckelshaus* (No. 84–1524 and consolidated cases, D.C. Circuit).
5. See, e.g., Burns, M., "The Law and Radioactive Wastes: A Shift to the States," 15 *Chemtech* 479 at 480 (1985).
6. South Carolina, Washington, and Nevada.
7. See P.L. 96–573.
8. P.L. 99–240.
9. P.L. 99–240, Title I, §5.
10. P.L. 99–240, Section 10(a).
11. P.L. 99–240, Section 10(b).

12. See the *Congressional Record*, December 9 and 19, 1985. For a discussion of the deferral of joint NRC and EPA regulations, see the *Congressional Record*, December 19, 1985, at H13076.

13. See, e.g., Scroggin, D. G., "Assessment of Toxic Risk Is Key to Agency Regs," *Legal Times*, November 12, 1984, p. 22.

14. Clean Air Act, Sections 112, 122(a); 42 U.S.C. §§7412, 7422(a). See, e.g., Industrial Union Dept., AFL-CIO v. American Petroleum Institute et al., 448 U.S. 607 (1980) (the "OSHA Benzene Case"). But see "Comments on the Proposed Withdrawal of Proposed Standards for Benzene Emissions" (Washington, DC: National Resources Defense Council, April 13, 1984).

15. OSHA Benzene Case at 652–3.

16. See, e.g., 49 Fed. Reg. 46294 (November 23, 1984).

17. RCRA, Section 1004(5)(B); 42 U.S.C. §6903(5)(B).

18. CERCLA, §101(24); 42 U.S.C. §§9601 et seq.

19. Clean Water Act, §307(a)(4); 33 U.S.C. §1317(a)(4).

20. FIFRA, §2(bb); 7 U.S.C. §136(bb).

21. TSCA, §#→§)(1)(A)(i); 15 U.S.C. #2603(a)(1)(A)(i).

22. SDWA, B1412(e)(3)(D); 42 U.S.C. B 300g-l(e)(3)(D).

23. 49 Fed. Reg. 46294–46331 (November 23, 1984) and 50 Fed. Reg. 1170 (January 9, 1985).

24. National Research Council, *Risk Assessment in the Federal Government: Managing the Process* (Washington, DC: National Academy Press, 1983).

25. *Risk Assessment and Management: Framework for Decision Making* (Washington, DC: U.S. Environmental Protection Agency, December 1984).

26. Ruckelshaus, W.D., "Science, Risk and Public Policy" (speech before the National Academy of Sciences, June 22, 1983), p. 11.

27. See, e.g., Ruckelshaus, W. D., "Risk, Science, and Democracy," 1 *Issues in Science and Technology* 19 (Spring 1985).

28. "Report by the SAB Carcinogenicity Guidelines Review Group" (Washington, DC: Science Advisory Board, U.S. Environmental Protection Agency, June 1985).

29. Because the generic guidelines will set an influential precedent, they can also be expected to affect the standards for adequacy of risk assessments used by other government agenices, such as the Occupational Health and Safety Administration (OSHA), the Food and Drug Administration (FDA), the Consumer Product Safety Commission (CPSC), and the Department of Health and Human Services (HHS).

30. E.O. 12291, 46 Fed. Reg. 13193 (1981).

31. Letter of August 12, 1986, from Wendy Gramm (OMB) to EPA Administrator Lee Thomas. Without such explicit disclosure of uncertainty information in risk assessments, Ms. Gramm expressed OMB's concern that the characterization of the health risk would not be "realistic."

32. "OSTP Comments on EPA Guidelines for Carcinogenic Risk Assessment," White House Office of Science and Technology Policy, Washington, DC, July 15, 1986, at 6.

33. Idem at 5.

34. Letter of July 25, 1986, from David T. Kingsbury (Asst. Director of NSF) to James Kamihachi (Information and Regulatory Affairs, OMB) at 2.

35. 51 Fed. Reg. 88992 (September 24, 1986).

36. There is a broad range of disagreement and controversy among scientists regarding the risk of human exposure to toxic and cancer-causing substances. Diet and lifestyle have been shown to be strongly correlated to incidences of cancer, whereas so far there has been no similar correlation with exposure to manmade chemicals. Nevertheless, the focus of the environmental statutes and of governmental efforts to protect public health have focused almost entirely on exposure to manmade substances and to industrial emissions. See Ames, B. N., "Dietary Carcinogens and Anticarcinogens," 221 *Science* 1256 (September 23, 1983); "Cancer and Diet," Letters Section, 224 *Science* 658 (May 12, 1984); and Scroggin, D. G., "Study May Be Used To Challenge Environmental Regs," *Legal Times*, October 31, 1983, p. 15.

37. "Benzene Emissions from Maleic Anhydride Plants, Ethylbenzene/Styrene Plants, and Benzene Storage Vessels: Proposed Withdrawal of Proposed Standards," 49 Fed. Reg. 8386 (March 6, 1984).

38. "Comments on the Proposed Withdrawal of Proposed Standards for Benzene Emissions" (Washington, DC: National Resources Defense Council, April 13, 1984), p. 13.

39. See, e.g., Wilson, E.B., *Introduction to Scientific Research* (New York: McGraw-Hill Book Company, 1952).

40. Courts have held that an agency's action may be invalidated as arbitrary and capricious when the agency relies upon an environmental impact statement that fails to disclose the full range of alternatives and environmental impacts. See, e.g., *Sierra Club* v. *U.S. Army Corps of Engineers*, 701 F.2d 1011 (2d Cir. 1983); *Environmental Defense Fund* v. *Froehlike*, 473 F.2d 346 (8th Cir. 1972); *Monroe County Conservation Council* v. *Volpe*, 472 F.2d 693 (2d Cir. 1972).

41. *Risk Assessment and Management: Framework for Decision Making* (Washington, DC: U.S. Environmental Protection Agency, December 1984).

42. *Environmental Defense Fund* v. *Ruckelshaus* (D.C. Cir., No. 841524) (radionuclides); *see also Natural Resources Defense Council* v. *Thomas* (D.C. Cir., No. 84-1387) (benzene), and *NRDC* v. *EPA*, 804 F.2d 710 (D.C. Cir. 1986), and (D.C. Cir.). The author's firm is counsel to parties in these proceedings. Although Administrator Ruckelshaus withdrew the controversial proposed radionuclides regulations in October 1984, based on his determination that the health risks were not significant, a federal district court nevertheless mandated that the agency promulgate final standards. In response, EPA promulgated emissions standards, but consistent with its earlier determination that health risks from current exposures were not significant, limited radionuclide emissions to current levels (50 Fed. Reg. 5190 [February 6, 1985]). For a more detailed history of the radionuclides

rule making, see Scroggin, D. G., "The Interaction of Science, Policy, and the Law in Agency Use of Risk Assessments for the Regulation of Carcinogens," 1 *Haz. Waste* 363–375 (1984) and Scroggin, D.G., "New Guidelines Face Complicated Problem of Risk Evaluation," *Legal Times* 20 (February 23, 1987).

43. Alm, A.L., "Managing Environmental Risks," 4 *Environ. Forum* 12, 13 (May 1985).
44. Stewart, R.B., "The Role of the Courts in Risk Management" (paper presented at the American Bar Association Fourteenth Annual Conference on the Environment, May 18, 1985).
45. Yuhnke, R.E. (attorney for the Environmental Defense Fund [EDF]), EDF news release, October 23, 1984.
46. "Report on the Scientific Basis of EPA's Proposed National Emission Standards for Hazardous Air Pollutants for Radionuclides" (Washington, DC: Subcommittee on Risk Assessment for Radionuclides, Science Advisory Board, U.S. Environmental Protection Agency, August 17, 1984), p. 32.
47. Idem, p. 3.
48. The New Jersey "Worker and Community Right To Know Act," N.J.S.A. 34:5A-l et seq., covers both manufacturing and nonmanufacturing sectors. This act was recently found to be partially preempted by OSHA's chemical "Hazard Communication Standard," 29 C.F.R. 1910.1000(b)(l). See *New Jersey Chamber of Commerce* v. *Robert E. Hughey*, No. 843255 (D.N.J. January 3, 1985).
49. See, e.g., H.R. 2576, the "Toxic Release Control Act of 1985," introduced by Reps. Wirth, Waxman, and Florio on May 22, 1985.
50. See, e.g., Ruckelshaus, W.D., "Risk, Science, and Democracy," 1 *Issues in Sci. & Tech.* 19 (Spring 1985).
51. For example, Rep. James J. Florio's "Chemical Manufacturing Safety Act of 1985," introduced in the 99th Congress, contains such a federal cause of action.
52. For example, a federal court recently reviewed the scientific risk assessments evidence and held that Bendectin was not the cause of birth defects alleged by plaintiffs. See *In re Richardson-Merrell, Inc. "Bendectin" Products Liability Litigation* (No. II), 606 F. Supp 715 (Judicial Panel on Multidistrict Litigation, April 11, 1985).
53. "The Truth about Agent Orange," *N.Y. Times*, Aug. 13, 1984, p. A22.
54. *Ayers* v. *Township of Jackson*, No. A-210383-T3, June 4, 1985 (see *Legal Times*, June 10, 1985, p. 4).

CHAPTER 9

Making the World Safe for Chicken Little, or the Risks of Risk Aversion

Letty G. Lutzker

Chicken Little and her friends, you will recall, formed a public interest research group to demand government protection from what they perceived, on the basis of misinterpreted data and unsubstantiated hearsay, to be a danger. In their preoccupation with minimizing a misidentified risk, they were unable to recognize the real threat posed by Foxy Loxy.

Many of those who as children laughed at Chicken Little's silliness and her friends' gullibility have grown into the adults about whom political scientist Aaron Wildavsky has said,

> How extraordinary! The richest, longest-lived, best-protected, most resourceful civilization, with the highest degree of insight into its own technology, is on the way to becoming the most frightened. Has there ever been, one wonders, a society that produced more uncertainty more often about everyday life? [Uncertainty about] . . . the land we live on, the water we drink, the air we breathe, the food we eat, the energy that supports us. Chicken Little is alive and well in America.[1]

A recent poll indicates that the majority of Americans perceive themselves exposed to more risk now than in times past.[2] Yet life expectancy is greater than at any previous time, has increased approximately 30 years since the turn of the century, and increased at a faster rate in the last 30 years than before. Some of this can be attributed to improved obstetrical practices that have lowered infant and maternal mortality, and elimination of many epidemic diseases that formerly killed young people, but there has also been an apparent improvement in older people's health as well. According to the statistics maintained by a large pension fund, 54% of men and 65% of women reaching the age of 65 in 1955 could expect to

Low-Level Radioactive Waste Regulation: Science, Politics, and Fear, Michael E. Burns, Ed., © 1988 Lewis Publishers, Inc., Chelsea, Michigan—Printed in USA.

live to age 80. Those percentages had risen to 67% and 78%, respectively by 1985. The percentage reaching age 65 in 1985 that could expect to live to age 85 was nearly that which could have expected to live to age 80 in 1955.[3]

Yet preoccupation with risk has increased. Risk analysis is now a recognized academic discipline complete with journals, societies, experts, equations, and jargon. At its heart are certain commonsense principles requiring that the existence of a risk be shown by some objective criteria, that benefit gained for risk assumed be considered, that risks of alternatives for the activity under review be examined and compared, and that the risk of eliminating the risk under review be evaluated. The Chicken Little school repudiates these principles in favor of a subjective approach that considers *fear* of an effect evidence of risk and ignores or minimizes the importance of benefits or alternative risks. The result is a thoroughly inconsistent approach that at the very least does not decrease the overall level of risk, and more often increases it by diverting resources from more beneficial efforts or exposing the public to new dangers.

CALCULATING RISKS

In 1978, the Food and Drug Administration (FDA) issued a warning about and considered banning nitrites as preservatives because in combination with amines in meat they form nitrosamines, substances shown to be carcinogenic in every animal tested – if, of course, the animal receives enough of them.[4] The original reason for adding nitrites in the first place was forgotten: it prevents botulism. Meanwhile, bacteria in mouths and intestines around the country were busily converting nitrates in potatoes to nitrites and combining them with amines in fish to produce nitrosamines, in open, or more accurately covert, defiance of the FDA. A real benefit was almost ignored in favor of a theoretical risk.

An inevitable consequence of water chlorination is the creation of a small amount of chloroform, in a concentration of approximately 100 ppb.[5] There is concern today that chloroform is carcinogenic, but little public recognition that cholera and typhus epidemics periodically ravaged cities before public water supplies were chlorinated. This same chloroform concentration was obviously less dangerous a few years ago, when 0.1 ppm was too small to measure.

Whooping cough, in the days before the pertussis vaccine virtually eradicated it, killed one in every 3000 children who contracted the disease. Vaccination carries a probability of 1 in 100,000 to 300,000 of causing neurological damage to the recipient. In the last few years, many British and American mothers with the best interests of their children at heart have refused to have those children vaccinated. Escalating "pain and suffering" awards in liability suits against manufacturers brought on behalf of the few unfortunate victims have driven all but one supplier from the business. The per-dose cost has risen from $0.11 to $3.11 in New York during the last three years.[6] Those mothers sane enough to wish their children vaccinated will inevitably pay more for the privilege, which is to say that some will be denied its benefit because of inability to pay more. "Most people would regard a ban on seat belts, based on occasional adverse events and in disregard of their safety benefits, as stupidity of the highest order. . . . Why, then, do they accept comparable actions in other contexts?" asks Chris Whipple, past president of the Society for Risk Analysis.[7]

Because of enormous litigation costs, even when the manufacturer won the suits brought against it, no intrauterine device is now manufactured in the United States. Women may now choose among contraceptive measures that are less effective, less personally appealing, less safe, or more radical and irreversible. Or they may forego contraception and opt for abortion or unwanted pregnancies, either of which is riskier than the IUD. In a recent court decision, $5,000,000 in damages was awarded to a woman who had used spermicidal jelly and subsequently borne a deformed child. The judge, reviewing the evidence obtained in a large study showing no increased incidence of anomalies in users of spermicides, said that evidence was irrelevant to the decision.[8] At the same time obstetricians, like physicians in other frequently sued specialties, are retiring early to avoid the threat of malpractice suits brought by patients who require absolute insulation from any chance of adverse outcome, an attitude fostered by "consumerists." Presumably a deficiency of goods and services for consumers to consume will decrease consumer risk.

In a panic about a supposedly impending swine flu epidemic in 1976, public health officials administered a vaccine to millions, including many for whom flu is usually not serious. An epidemic of Guillain-Barré syndrome, a debilitating and sometimes permanently disabling neurological disorder, ensued.[7]

A public that is fearful of adding pesticides to preserve agricultural products applauds the development of pest-resistant strains that make pesticides unnecessary. Plants protect themselves from being eaten by pests by manufacturing their own toxins, which cannot be washed off.[7]

Cyclamates were banned because massive doses caused bladder cancer in rats. Saccharin remains legal, although in Canada it was saccharins that were prohibited for the same reason, while cyclamates are permitted.[7] Even more illogically, the pesticide EDB was banned after administration to rats of 10,000 to 50,000 times the usual dietary content caused stomach cancer, although the incidence of that malignancy has been steadily declining in the United States for 20 years. When this absurdity, in a world where millions starve while one-third of the food produced rots before it can be distributed, was pointed out to the editor of the *New England Journal of Medicine*,[9] Massachusetts health officials were not embarrassed to respond in print that "overall cancer rates are high and . . . rising . . . lung cancer rates . . . especially. . . .Under these circumstances, the data from animals on the carcinogenicity of EDB appear especially troublesome for human beings."[10] Lung cancer rates are indeed rising, but ingested EDB cannot cause lung cancer, the rise in which is more clearly associated with cigarette smoking, which these health guardians did not mention.

THE FEAR OF POLLUTANTS

The near-impossibility of siting any waste disposal facility exemplifies many of the problems of risk aversion practiced subjectively. Since some time around the early 1960s, the notion has grown popular that the earth is a fragile ecosystem in imminent danger of destruction by man's industrial activities. Fear of pollutants spread, despite a lack of independently corroborated evidence that pollution-related disease and destruction were actually occurring more extensively than before. In this apocalyptic environmental nightmare, evil humors emanating from man's industrial wastes are killing directly, by poisoning air and water, and indirectly, by mangling genes that will turn future generations into monsters. With the end of the world so imminent, small wonder that sober thinking became so difficult.

"Suddenly," says philosopher and ethicist Margaret Maxey, "the bene-ficiaries of Western industrialized society are conscience-stricken by the discovery that there is no longer an 'away' to throw things into."[11] Eradi-cation of the plagues and epidemics that ravaged the civilized world in less sophisticated times should have taught the lesson that wastes, though dangerous if left lying around unmanaged, especially upstream, could be safely handled if properly handled. Instead of practicing "source reduc-tion" by ceasing to eat, or quaking before the invisible specter of disease-carrying "night humors," some of our forebears fortunately invented sewage isolation and treatment methods instead. Now such plagues and epidemics occur, by and large, only among peoples privileged to live in more "natural" pretechnological circumstances, the inroads of modernity on which are bemoaned in public television documentaries.

The earliest and still loudest opposition was to radioactive waste dis-posal facilities, radiation holding special terror for many. It is not only invisible, impalpable, and inaudible, like the miasmas against which our less sophisticated predecessors closed their shutters, but odorless as well. Although its physical characteristics make it actually easier to track in minuscule amounts, it has come to symbolize all that is insidiously threatening.

Looked at in isolation from the universe of which they are a part, radiation risks might indeed arouse concern. Experimental evidence abounds to show that *large* doses of ionizing radiation can be carcinogenic, teratogenic, and mutagenic. But cancer, malformed fetuses, and genetic mutation occurred before man began to use the atom for his own purposes and must be accepted, albeit sadly, as afflictions the root causes of which are not yet known but that clearly include factors other than the radiation to which the human race has been subjected as occupants of the physical universe. In response to popular fears, however, the Environmental Protection Agency has established strict limits on flight crew exposure to the negligible radiation of radiopharmaceutical packages, ignoring in their calculations atmospheric doses, which were higher to begin with and which have actually increased in recent years as higher-altitude, higher-latitude flights have become more common.[12] The lifetime risks of flight crews have probably not been affected one way or the other, but the costs of radiopharmaceutical transport have certainly been increased by this ideological distinction between natural and manmade sources of radiation.

RADIATION AND CANCER

If we calculate the *incremental* risk of a particular amount of radiation to decide whether to fear it, the result is startlingly insignificant. The average American has roughly a 35% chance of contracting cancer in his lifetime, radiation notwithstanding, and approximately a 20% chance of dying from it. That is to say, in a population of 1,000,000 persons, 350,000 will develop a cancer and 200,000 of those cancers will cause death. The addition of one rad to that entire population, an amount 5–10 times the natural background in this country and 50–100 times that which would be added to the annual population dose if all electricity were nuclear, can be theoretically estimated (using the "straight-line hypothesis" that will be further discussed later) to increase these probabilities by 0.01%; 200,100 of the 1,000,000 individuals will die of cancer. This is certainly within the usual fluctuations in cancer occurrences and deaths from year to year. Genetic mutations occur in approximately 10.7% of live births; application of the above-mentioned rad to 1,000,000 fetuses would theoretically raise that likelihood to 10.7075%.

Any gambler would recognize a change in odds from 1 in 5000 to 1 in

4997, or 1 in 9345 to 1 in 9339, as a nonexistent change. But not all citizens play the horses, and a 40-point fall in Scholastic Aptitude Test (SAT) mathematical scores in the last 20 years suggests that for many Americans elementary arithmetic may not provide a comprehensible method for determining significance. It must be remembered also that risks are not necessarily incremental. Perhaps women having regular mammographic examinations are exposing themselves to a one-in-several-thousand chance of developing a radiation-induced cancer. But since one in every 11 women in the United States develops breast cancer, and the mortality of small lesions detected early is significantly lower than those found later, radiation-phobia may shorten life more than radiation.

Fluctuations in the annual cancer occurrence rates are larger than the postulated additional effects of radiation. It is also often forgotten that 100 "additional" cancer deaths per million does not mean 100 additional deaths, the population of 1,000,000 being presumably mortal. Cancer deaths as a percentage of total deaths are in fact rising, because of significant decreases in death rates from other causes in recent years, most notably cardiovascular disease, still the largest "killer" of old people. It is not abundantly clear why death from cancer should be considered more offensive than death from other causes, unless it is untimely, as every single one of the 14,000 automobile-related deaths that would occur in that population of 1,000,000 over its collective lifetime certainly would be. Cancer, on the other hand, is largely a disease of advancing age, more than 55% of cases occurring above the age of 65.[13] Age-adjusted cancer incidence rates have not changed during the more than five decades that extensive national records have been kept, despite the rampant advance of industrialization, but total cancer incidence has, as the population lives to be older. In less well-developed countries, where cancer prevention takes the form of dying from malnutrition, famine, or infection by the age of 40, cancer incidence and the ratio of cancer-related deaths to all causes of death are indeed lower than ours.

A frequently used justification for the blasé acceptance of traffic deaths is the fallacious distinction between voluntary and involuntary risk. Driving one's car is viewed as an act of free will for which the individual accepts responsibility, and potential radiation from power plants or waste sites is a risk imposed by outside sources. Of course, many employers, by requiring their employees to appear on time for their jobs, are obliging them to drive, and for the pedestrians and bicycle riders who figure in traffic mortality statistics the driver's so-called voluntary decision imposes a clearly involuntary danger. Chicken Little fears cancer more than automobiles, and her very fear actually makes cancer more dangerous, as author-humorist Garrison Keillor notes:

> Poor Pete. Cancer got him. He always knew it would and in his last years kept a desperate watch for the signs of it — the Seven Danger Signs was taped to his bathroom mirror — but without much hope: every day revealed a possible sign, something unusual, a little change of weight, a thickening, a slight lump, some soreness, a redness of the stool, a sore that was slow to heal (older guys heal slower) — then, that fateful Friday, he felt a definite lump on the back of his head and was dizzy and found blood on his toothbrush. Lois was off to clean the church and he panicked — jumped in the car in his pants and T-shirt — it was Dr. DeHaven's day off and besides, Dr. DeHaven didn't believe his cancer theory — so he headed for St. Cloud to a new doctor, and only a panicky man would have passed that semi the way they said he did, on a long right-hand curve going up the hill toward Avon, and there he met his end and found his peace in the grille of a gravel truck.[14]

THE ROLE OF GOVERNMENT

The ecological panic of the last 20 years included a cancerphobia the development of which was certainly abetted by, and probably at least in part created by, government officials claiming to have the population's welfare in mind. Anonymous pronouncements and unauthored press releases from government agencies such as the National Institutes of Occupational Safety and Health, National Cancer Institute (NCI), National Institutes of Environmental Health Studies (NIEHS), Occupational Safety and Health Administration (OSHA), and FDA have created a "cancer crisis" not borne out in any peer-reviewed epidemiological study from their own research groups or any other, and have initiated a frantic search for corporate carcinogens. A chief of laboratories at the NCI has averred that those who perform animal testing of industrial chemicals for carcinogenicity will be "bearing witness to mass murder."[4] OSHA has stated "if available evidence indicates that the chemical is not positive for carcinogenicity, OSHA's position is the same as if the chemical had never been tested."[4] Since OSHA tests only chemicals used in the workplace, it clearly has decided before the fact that industry can never be declared innocent of charges of causing illness and death for profit. When asked why the FDA did not issue warnings about natural carcinogens in food, a spokesman said, "Such warnings would be so numerous they would confuse the public . . . not promote informed consumer decisionmaking . . . not enhance public health."[4]

In a democracy, it is right and proper that public officials, elected or appointed, be responsive to the expressed concerns of the citizenry. As long as there are citizens who express unreasonable and inconsistent fears, there will be politicians who will respond to those fears and even exploit them. A former commissioner of FDA, which has issued public warnings about potential health threats of thousands of food additives, has stated, "It is clear . . . from the health regulatory laws and from judicial interpretation of those laws, that regulatory action to protect the public health from a perceived risk is appropriate, even when the perceived risk is based on a mixture of scientific fact, theory, and supposition,"[4] acknowledging that the purpose of government officials is to respond to popular concern rather than to protect the public from a real danger. As Massachusetts Congressman Edward Markey succinctly stated at a discussion on low-level radioactive waste disposal, "Perception is reality."[15] Thus spake Chicken Little.

Because of the known relationship between large doses of radiation and cancer development, the straight-line hypothesis was developed as an operational tool to permit definition of conservatively safe standards in the absence of firm knowledge of where a threshold for effect lay. Increases in cancer incidence have been convincingly documented in human populations exposed to 50 rem or more, and then only if the dose was received all at once to the whole body.[16] Much higher doses are required if only single organs or body parts are radiated.[17,18] Popular mythology aside, there is no evidence of "cumulative" effects of low doses; to the contrary, experimental evidence shows the effect of any particular radiation dose to be less if delivered over a longer time than a shorter. Deleterious effects of low doses (less than 10 rem — total body) have never been demonstrated experimentally or epidemiologically, and are unlikely ever to be, since they are clearly very small, if they exist at all, in relation to the "natural" incidence of cancer, genetic disease, and congenital malformation. (There is abundant experimental evidence showing a beneficial effect of low-level radiation exposure on a variety of plant and animal systems,[19] comparable to the positive effects of small amounts of many substances that can be harmful at larger levels. Many in the scientific world who know of this evidence, not acknowledged by the science watchers of the press, are embarrassed to speak of it lest they be laughed off the scene.) Yet the straight-line hypothesis has been widely accepted, even among many scientists who should know better, as proof itself that there is no threshold and therefore no definitely "safe" dose. Lauriston Taylor, past president of the National Council on Radiation Protection, urges that we "stop arguing about the people who are being injured by exposure to radiation at the levels far below those where any effects can be found despite over 40 years of trying to find them. The theories about people being injured have still not led us to the demonstration of injury and, though considered as facts by some, must only be looked upon as figments of the imagination."[20]

NUCLEAR ACCIDENTS

Another nuclear-related figment is death from plutonium, usually described as the "deadliest substance known to man," when that phrase is not being applied to dioxin. Unlike dioxin, which must be eaten to damage human health (and eaten in greater quantities than, for example, cyanide, which is widely known to be deadly, is rarely described as the "deadliest," and is present in almonds), plutonium is relatively innocuous unless inhaled. It is so difficult to disperse in air, however, that it is not easily inhaled. Ralph Nader has claimed that a pound of plutonium dispersed in air and inhaled would kill 8,000,000 people. Dr. Bernard Cohen, a University of Pittsburgh nuclear physicist, suggests that if so dispersed it would more accurately kill 2,000,000, but more important,

1. It cannot be so dispersed before it would clump and fall to the ground where it would have to be eaten.
2. It is 5000 times less toxic when eaten than when inhaled.
3. Open-air bomb tests have already deposited 10,000 pounds of it into the atmosphere (one would think that 2-8 *million* extra deaths would have been noticed).
4. We manufacture annually enough chlorine, phosgene, ammonia, and hydrocyanide to kill many *billions* if these substances are dispersed to the same specifications.[21]

Toxicity, after all, is not the only factor that determines hazard. Concentration and availability must be considered for any meaningful risk determination. Electricity itself, which kills more than 1000 Americans annually,[21] is rarely referred to as "lethal."

With certain refreshing exceptions, the print and electronic media have preferred to reflect and intensify public fears than to educate. This bias was graphically illustrated in Dr. Cohen's review of *New York Times* coverage of accidental deaths in the four years preceding the Three Mile Island (TMI) accident.[21] There were an average of 120 write-ups per year on highway deaths, which kill 50,000 per year (one story per 417 deaths); 240 stories per year about industrial accidents, which kill 12,000 annually (one story per 240 fatalities); and 25 stories about suffocation, which kills 4500 (one story per 180 deaths). This trend toward more coverage of rarer events feeds the public's appetite for catastrophe, an appetite apparently infinite for fictitious catastrophes; there were 200 stories on the possibility of deaths related to nuclear energy, none of which had occurred during that time. After TMI, the number of nuclear scare stories increased although no one was killed, and every March since 1979 a spate of anniversary stories appear. Relatively less was written about the Mount St. Helens eruption, which spewed out approximately the same number of curies of radon as TMI did xenon, but radon is 1000 times more carcinogenic.

Since then, of course, approximately three dozen people have died as a result of the Chernobyl accident, some from the effects of trauma sustained in the explosion and fire, but many certainly because of radiation injury. These should properly be considered occupational deaths, 30 of which in any other occupation would not generally excite weeks of news coverage. An oil rig that killed 140 workers in 1984 went virtually unnoticed, as did the fiery deaths of hundreds of Mexicans in a natural gas explosion in December 1984. The land contaminated by the Chernobyl accident is comparable to, though smaller in extent than, the thousands of acres destroyed when the Grand Teton Dam burst in 1976.[22] The point is not that the occurrence of one kind of disaster legitimizes another, but rather that energy technologies have risks, and that, measured in terms of human and property costs, those of nonnuclear technologies are by no means negligible.

Released radioactivity blown around the world added to the excitement following the Chernobyl accident, of course. Concentrations of I-131 in drinking water in this country did not reach the minimum level determined by the Environmental Protection Agency to require protective action (15,000 pCi/L). It is interesting that daily consumption for a year of a quart of water contaminated to that Protective Action Guide level would deliver a thyroidal radiation dose approximately equal to that delivered by one medical radioiodine uptake test for thyroid function. Ten years of such consumption would deliver a thyroidal dose approximating that delivered by a scan done with I-131, a procedure shown to have caused no excess thyroid cancers in a study of several thousand patients.[23]

Although the media are the sources of scientific information for most of the American public, training in science is not a prerequisite for media science writing. But lack of scientific training does not explain the sensationalistic tone of most reporting of radiation issues, more especially in the electronic media than in print, nor does it explain the reluctance of the media to publicize the views of those scientists, the vast majority, who would dispute some of the factual errors or lurid claims made. An alarming article by a *New York Times* science writer purported to alert the public that radiation risks may be higher than previously thought.[24] Speculations about incompletely interpreted data and uncorroborated results of disputed small studies abound. Apparently able to divine results of research not yet completed, the author even makes the startling claim that ongoing studies will "almost certainly lead to the conclusion that radiation is a greater risk than most official estimates have indicated." The factual and interpretive errors were pointed out in a comprehensive critique by a group of radiation scientists whose credentials included serving on national and international radiation protection commissions; the corrections were not printed.[25]

Following the *Mont. Louis* shipwreck, both *Time* magazine[26] and *Newsweek*[27] carried stories filled with hints of ocean contamination and exploding uranium hexafluoride. I am sure I was not the only one to express, in writing, my dismay at the omission of some basic facts, such as the impossibility of a UF_6 explosion, and the distortion of others, such as the relative environmental and economic consequences of this incident compared to oil spills. Neither my nor anyone else's objections appeared in these magazines; one promised me to correct its files.

RISKS THAT REPLACE RISKS

A specific technological risk should not in fact be considered an incremental addition to total risk if the new activity displaces or replaces another activity that is riskier. The "falling skies" of nuclear energy production become "acorns" when we consider the consequences of the available alternatives to nuclear electricity production, which fall into two categories: (1) using less energy altogether, or (2) using other fuels to obtain the energy we need.

Our society is energy-intensive, and, romantic notions notwithstanding, we are more comfortable and healthier because of it. Although the links between economic well-being and longevity are not fully understood, it is evident that people live longer in the developed nations of the world than in the poorer ones, and that the correlation between socioeconomic status and longevity is in general positive. Even if per capita consumption remains stable, total energy use must increase if the population grows. If the economy is to expand, greater increases are required. The former contingency is inescapable and the latter presumably desirable. Conservation, a sensible practice because wasting any resource is stupid, is not an energy source per se and cannot alone meet the pressures of increased population. Should those who now have less of the goods and services that money can buy forego the option of ever having more? Are those who now benefit from our economy but who mouth notions of being less materialistic really prepared to agree to make do with less? There are likely to be nearly 30,000,000 more occupants of this country by the year 2000, including 8,000,000 more senior citizens (among whom I will count myself by 2007) who will be using up rather than generating income, requiring greater health care funding, and be unwilling to turn down the thermostat. Foxy Loxy lurks in the likelihood of the social unrest and disruption that occur in periods of scarcity when those who do not have realize that they will be unable ever to get, in the rationing of limited resources with resultant greater governmental intrusion into citizens' lives, and in a gradually declining living standard for all. These risks cannot be directly measured, of course, but they should be contemplated by any who advocate economic stagnation as desirable.

Energy needs will rise, in the developing world even more rapidly than in the technologically advanced countries. In the latter, that energy is more and more electrical. What are the risks of not developing nuclear energy to meet these needs?

One is the "great danger that if nuclear energy is not developed fast enough, wars may become possible for the simple reason of competition for oil and natural gas."[28] In the flush of abundant cheap oil, the American public has forgotten that the supply of oil is finite, that the easily tapped fields have been used, so the technical obstacles to obtaining the dwindling supply must increase and therefore the absolute cost will rise, and that more than 60% of the world's known reserves are in the Middle East. All of the ingredients for worldwide devastation, economic and military, are present. In the name of safety, antinuclear activists have seen to it that 12 years or more are necessary in the United States to build a nuclear power plant that can be constructed in six years in Japan or France, using technology pioneered in the United States. The time required for a presidential decision to protect the nation's oil supply by military means is measurable in minutes. The industrialized areas of the Northeast, Midwest, and Southwest have 80% of the nation's oil, coal, and nuclear plants. The latter are 100% utilized, while the former two are mostly idle, waiting to be brought on line when electricity demand outstrips supply, around the turn of the century, in a country where no major new capacity is being planned.[29] Oil prices, now low in large part, as the Organization of Petroleum Exporting Countries itself admits,[30] because of nuclear utilization, will have risen again by then. In 1984, before the current oil price drop, and after years of efforts at conservation, the U.S. trade deficit attributable to oil importation approximated that of imported textiles and automobiles *combined*.[31] Had we not had our nuclear capacity, that deficit would have been twice as high.

NUCLEAR POWER AND NUCLEAR WEAPONS

A confusion between megatons and megawatts has clouded energy policy deliberations. Fear of nuclear weapons proliferation led President

Carter to abandon the breeder reactor and spent-fuel recycling programs, which would have assured future generations of an uninterrupted
energy supply for thousands of years, not mere decades or centuries.
Plutonium, it was feared, could be diverted into weapons production.
This claim was believed despite the fact that no nation has as yet developed the bomb through its domestic nuclear electricity program, for a
variety of reasons, the major one being that it is the long way around to
get good weapons-grade plutonium. Terrorists would have a difficult job
stealing the stuff without damaging themselves under any circumstances,
but especially so if all enriched uranium production and reprocessing
were at least under U.S. control, rather than scattered around the world,
as it has become since the U.S. lost its credibility as a major supplier.
Thirty years ago, the United States supplied all of the enriched uranium
used for domestic energy production in the non-Soviet world; it now
supplies 35%.[32] With loss of that market went loss of leverage to convince governments interested in developing nuclear industries to sign
nonproliferation agreements. This country's "negative attitude toward
nuclear power . . . was seen as a challenge to other countries to achieve
energy independence."[33] It seems that "instead of reasonable precautions
against the nuclear sword, we have built an impregnable defense against
the nuclear plowshare."[34]

Coal burning, although widely recognized to have direct deleterious
health and environmental effects, does not possess the mystique of
nuclear fission. During an inversion that raised pollution levels in London in 1952, more than 3500 people—members of the public, not workers in the industry—died in one week and a few thousand additional
deaths occurred in the following weeks. Several smaller pollution epi-
sodes have been documented in New York City. It is estimated that
10,000–50,000 Americans die from pollution-related respiratory disease
each year. Acid rain is damaging vegetable and animal life in places far
from those receiving the benefit of the burning coal. Mining of coal is
more destructive to scenery and miners than is uranium mining. The
environmental and economic dislocations that will result from the
"greenhouse effect" of combustion are yet to be calculated, but it is
certain that future generations will pay whatever price there is to pay and
likely that they will not thank those who left them the legacy.

During the decade of energy awareness and increased conservation,
wood burning has increased dramatically. Wood use has certainly dis-
placed some oil use, and that is all to the good. If wood-burning stations
had been built instead of coal and nuclear plants during that decade, the
nation would have had to be deforested each year. Burning of wood
produces more pollutants than burning of coal, including potent carcinogens such as benzopyrene and other polycyclic aromatic hydrocarbons,

and poisonous carbon monoxide. The relative public health hazard of wood burning has been estimated as 30–100 times that of coal burning.[35,36]

The descriptive term *soft* when applied to energy production suggests benignity, and the term *renewable* implies a limitless supply. The electricity supplied by one conventional large electric plant would require 150 square miles of windmills and raise problems of noise pollution, land use, construction materials, and what to do when the wind is not blowing. The electricity use *increases* during the 10 years following the oil embargo, supplied by 100 new coal and nuclear plants, would have required construction of 1.5 million windmills covering 15,000 square miles. Where would the snail darter have gone then?

Solar units, even should the costs become competitive with conventional technology, cannot be of much use in Boston in February; necessary backup systems are expensive because they must be left idle until needed but paid for anyway. The cadmium, aluminum, copper, and other minerals necessary for solar equipment are, like the components of windmills, not renewable. Refinement of the ores in which the minerals used in solar technology are found frees pollutants (such as arsenic, which, having no half-life, is toxic forever) to enter the water supply.

IDEOLOGY

"A commitment to 'soft energy' . . . [is] . . . a romantic notion that seduces many individuals looking for an anticorporate, antigrowth rationale."[37] Individuals and groups with a variety of concerns and preoccupations — some concrete, some political, some mystical — could unite in opposing nuclear power plants or waste disposal facility development. In a risk-conscious society, radiation-phobia dovetailed nicely with cancerphobia and concern for the genetic integrity of future generations, and was associated in many minds with fear of a nuclear war's devastation. The seemingly specific debate about risks of technologies or waste disposal techniques masks a battle between warring moralities of which many of the participants themselves are unaware.

One wonders, with Aaron Wildavsky, "Is there something new in our environment or in our social relations?"[1] William Hendee, past president of the Society of Nuclear Medicine, acknowledges a darker side to the pastoral vision, stating that "many of us in nuclear medicine recognize the antinuclear movement for what it is, against big business, government, and the present social order."[38] Those "who advocate conversion to a solar-energy economy, coupled with the abandonment of currently available energy sources, are in fact proposing to change American society without explicitly indicating their intent."[39] Such utopians are convinced, like millennarians of the past, that man could be perfectible if corrupting influences were removed from society. Today's devil is large-scale corporate capitalism, frontal political attacks on which have failed in the past. By more subtly promoting distrust of all social, political, economic, and scientific institutions; by exploiting the phobic mentality that grips our risk-conscious society; by appealing to universal middle-class concerns for cleanliness and creating a better world for one's children, today's utopians have attracted the support of many who might not want their lives radically altered if they really thought about it. The devil cannot be radically exorcised, but it can be so effectively slowed, para-lyzed, or blocked that it becomes discredited as a viable economic option.[40]

Secure in possession of the leisure time, material benefits, and greater health that scientific discoveries and technological applications have pro-vided through a profit-driven system, the spoiled brats of the affluent society wish to "have their capitalistic cake while eating their capital-ists."[4] They can afford to be elitist, freely consuming the goods their much-maligned system produces while intoning solemnly that people should make do with fewer material things, perhaps meaning other peo-ple. They reveal their hypocrisy when they piously protect their progeny from the alleged threat posed by new U.S.nuclear plants while advocat-ing purchase of Canadian nuclear power, or when they oppose the Dickey-Lincoln hydropower project that would have flooded 140 square miles of Maine while they advocate buying Canadian hydropower pro-duced by the 4400-square-mile James Bay installation. The ". . . environ-mentalists' attitude fails to show consistent concern for the land and for the atmosphere but does show consistent hostility toward large corpora-tions, especially toward producers of energy."[41]

The world, after all, cannot be made safe for Chicken Little, who cannot distinguish between the reality of perceptions and the perception of reality. Having "nothing to fear (but a few zillion things),"[42] she is in a panic-stricken state in which it is impossible to exercise common sense. Distressed that science and technology are imperfect and that the benefits they confer are not free, she is willing to throw out the baby with the bathwater. There is no logical basis or practical use for the assumption that the effects of massive amounts of a substance on any animal can be extrapolated to the effects of small amounts on human beings. Nor does it make sense to consider only proof of danger as definitive, but demon-stration of safety as evidence of insufficient testing. Yet with chemical carcinogens as with radiation, this is the standard that Chicken Little and her official protectors, unwilling to be reassured by evidence, require. Critics of the lack of objective evidence supporting governmental agency decisions are answered by statements like one from the Toxic Substances

Strategy Committee, which reported that "proponents [of making regulatory decisionmaking more scientific] favor standardization of regulatory phases, clear separation of facts from value judgments, and organizational separation of scientists from policymakers. Although theoretically neat, TSSC does not consider it realistic. . . . Rather, flexibility in the decisionmaking process should be preserved without making artificial distinctions between science and policy."[4]

"Fear of the unknown is an ancient human bugaboo. Fear of the known and useful is modern American."[43] If Chicken Little continues to prefer slogans to facts and subjective impressions to objective data, to accept social paralysis for fear of what is unknown rather to encourage deliberate action based on what experience and investigation have taught, the sky might just fall.

REFERENCES

1. Wildavsky, A., "No Risk is the Highest Risk of All," *Am. Scientist* 67:32-37 (January-February 1979).
2. "How Much Risk? An Evaluation of Public Attitudes," *Pub. Opin.*, February-March 1986.
3. "More TIAA-CREF Retirees Now Live into Their 80s," *The Participant*, June 1986.
4. Efron, E., *The Apocalyptics: How Environmental Politics Controls What We Know about Cancer* (New York: Simon & Schuster, 1984).
5. Wilson, R., "Analyzing the Risks of Everyday Life," *Technology Rev.* (February 1979), 41-46.
6. Axelrod, D. (New York state commissioner of health), newsletter to the physicians of New York state, November 1986.
7. Whipple, C., "Redistributing Risk," *Regulation*, May-June 1985.
8. Miller, J. L., and Alexander, D., "Teratogens or 'Litogens,' " *New England J. Med.* 315(19):1234-1236 (November 6, 1986).
9. Stare, F. J., *New England J. Med.* 310:1387 (May 24, 1984).
10. Havas, S., and Walker, B., Jr., New England J. Med. 310:1387-1388 (May 24, 1984).
11. Maxey, M., editorial, *Houston Chronicle*, July 18, 1986.
12. Bramlitt, E. T., "Commercial Aviation Crewmember Radiation Doses," *Health Physics* (November 1985).
13. "Cancer Statistics 1986," *Ca: A Cancer Journal for Clinicians* 36(1) (January-February 1986).
14. Keillor, G., *Lake Wobegon Days* (New York: Viking, 1985), pp. 192-193.
15. Markey, E., comments at combined New England-Greater New York annual meeting, Society of Nuclear Medicine, October 1986.
16. Kohn, H. I., and Fry, R. J. M., "Radiation Carcinogenesis," *New England J. Med.* 310(8):504-511 (February 23, 1984).

17. Loken, M. K., "Low Level Radiation: Biological Effects," *CRC Crit. Revs. in Diagnostic Imaging* 19(3):175–202.
18. Webster, E. W., "On the Question of Cancer Induction by Small X-Ray Doses," *Am. J. Roent.* 137:647–666 (October 1981).
19. Luckey, T. D., *Hormesis with Ionizing Radiation* (Boca Raton, FL: CRC Press, Inc., 1980).
20. Wagner, H. N., "Radiation: The Risks and the Benefits," *Am. J. Roent.* 140:595–603 (March 1983).
21. Cohen, B. L., *Before It's Too Late: A Scientist's Case for Nuclear Energy* (New York: Plenum Press, 1983).
22. Hosford, A., "Technology Risks," computer search by AIF staff, June 1986.
23. Holm, L. E., Lundell, G., and Waslinder, G., "The Incidence of Malignant Thyroid Tumors in Humans after Exposure to Diagnostic Doses of I-131: a Retrospective Cohort Study," *J. Nat. Cancer Inst.* 64:1055–1059 (1980).
24. Boffey, P. M., "Radiation Risks Higher Than Thought," *N. Y. Times,* July 26, 1983.
25. Brill, A. B., personal communication, 1983.
26. Angler, N., "A Shipwreck Sends a Warning," *Time*, September 10, 1984.
27. Canine, C., "Tracking a Nuclear Near Miss," *Newsweek*, September 10, 1984.
28. Bodansky, D., "Risk Assessment and Nuclear Power," *J. Contemp. Studies* 5(1):527 (Winter 1982).
29. Science Concepts, Inc., "Return of the Age of Oil," report for the U.S. Committee for Energy Awareness (November 1985).
30. Stauffer, T., "OPEC Worries about Nuclear," *Energy Daily* 13(235):11 (December 11, 1985).
31. U.S. Department of Commerce figures, tabulated in "Information about Energy Americans Can Count On," U.S. Committee for Energy Awareness (1985).
32. Starr, C., "Uranium Power and Horizontal Proliferation of Nuclear Weapons," *Science* 224:952–957 (June 1, 1984).
33. Shultz, G., "Preventing the Proliferation of Nuclear Weapons," address to the U.N. Association of the USA (November 1, 1984).
34. Dodson, O., *North Louisiana Bus. J.,* April 21-May 21, 1984.
35. Science Concepts, Inc., "Wood," in "Energy Issues in Perspective" series, November 1982.
36. Travis, C. C., Etnier, E. L., and Mayer, R., "Health Risks of Environmental Wood Heat," *Environ. Mgmt.* 9(3) (1985).
37. Walske, C., and Dobkin, R. A., "The Nuclear Controversy," *Nuclear Activist* 15-18 (January 1984).
38. Hendee, W. R., "Nuclear Medicine and the Anti-Nukes," *Applied Radiology,* May-June 1982.
39. Wolfe, B., "Power to the People: The Nuclear Debate Isn't Really about Energy," *L.A. Times,* November 22, 1981.

40. Isaac, R. J., and Isaac, E., *The Coercive Utopians* (Chicago: Regnery Gateway, Inc., 1983).
41. Gilman, L., "Environmentalists' Two Standards," *N.Y. Times,* October 24, 1984.
42. Allman, W. F., "We Have Nothing to Fear (but a Few Zillion Things)," *Science '85,* October 1985.
43. Tyrrell, R. E., "The Anti-Nuke Priesthood," *Wash. Post,* July 29, 1985: "little to do with science and much to do with faith."

CHAPTER 10

The Low-Level Radioactive Waste Crisis: Is More Citizen Participation the Answer?

Richard J. Bord

INTRODUCTION

The goals of the 1980 Low-Level Radioactive Waste Policy Act have not been met. That act stipulated that regional disposal sites were to be established by 1986. To date, no new sites have been established and none are anywhere near the construction phase. Congress, responding to the existing impasse, has extended the deadline to the end of 1992 with the passage of the Low-Level Radioactive Waste Policy Amendments Act. The reasons for the impasse are no mystery: local intransigence regarding waste of any kind, public fears of radiation hazards, and politicians' anxieties about their constituents' fears. The focus of this paper is the viability of ongoing attempts to overcome public intransigence in the case of disposal siting for low-level radioactive waste (LLRW).

Daniel Bell asserts correctly that the United States has undergone a participation revolution, characterized by the belief that "people ought to be able to affect the decisions that control their lives."[1] Nelkin focuses this theme in terms of citizen response to risky technologies.[2] The public increasingly feels free to challenge issues formerly reserved to those with scientific or technical expertise, and the courts have put teeth into those challenges. The dominant problem now facing those responsible for implementing many technical decisions has become the promotion of public acceptance. Public intervention into decisions concerning risky technologies is generally viewed positively in light of our democratic norms and the ample history of corporate and government insensitivity to issues of public health and safety.

Low-Level Radioactive Waste Regulation: Science, Politics, and Fear, Michael E. Burns, Ed., © 1988 Lewis Publishers, Inc., Chelsea, Michigan—Printed in USA.

There are, however, reasons to be cautious about predicting the continuing vigor of the participation revolution, at least relative to certain risky technologies. It has become virtually impossible to establish new toxic chemical, radioactive, and, in some cases, solid waste disposal facilities.[3-7] Public reaction to the possibility that these substances will be introduced locally typically generates an adamant "not in my backyard" response. An intransigent public creates interminable delays, which often result in the scuttling of the project due to a loss of cost-effectiveness. Actual bargaining and negotiation that produce a mutually satisfactory decision are increasingly rare. One EPA official states the problem succinctly: "With darn few exceptions, public participation has become a stonewall opposition to siting."[3]

Public unwillingness to accept the burden of risky wastes has forced public participation programs to evolve in the direction of granting more and more concessions to the local public in the hope of winning their acceptance. However, the elaboration of public participation programs has not been accompanied by a proportional increase in public acceptance and cooperation. There may be risky technologies that the public is simply unwilling to tolerate. The case of low-level radioactive waste is used here as an example of a risky technology which may have pushed public participation programs to their logical limits without a forthcoming solution.

In this discussion the term "public opposition" encompasses a number of potential sources of dissent. Communities targeted for waste disposal sites tend to evolve local protest groups.[8] However, fear of radiation hazards is so widespread that statewide protest may mobilize even in absence of local opposition.[9] Politicians, viewing identification with radioactive waste disposal issues as a no-win situation, can be expected to be protective of their own states and districts. Finally, a nationwide network of antinuclear organizations can be expected to respond negatively to any attempt to site radioactive waste.

This paper has two parts: the first deals with the public-fear idiosyncrasies of the low-level radioactive waste issue; the second details the evolution of public participation programs under the press of an increasingly intransigent public. That analysis leads to a prognosis about the probable efficacy of future attempts to elicit public cooperation, using data from a Pennsylvania statewide survey to punctuate the arguments.

LOW-LEVEL RADIOACTIVE WASTE SITING: A "SPECIAL" PROBLEM

Public Fear: Rational or Irrational (or Does it Make Any Difference?)

In an analysis of public reactions to solid waste, Bealer and his colleagues note that there are deep cultural roots to a general dislike of, and disgust toward, any kind of waste.[4] Waste has nothing to recommend it. Its introduction into a community is viewed as a stigma that leads to esthetic and other penalties for local residents. Waste carrying the label "radioactive" shares those negative associations but also bears an added burden: intense public fear. As Freudenburg argues, nuclear materials inspire "special" reactions.[10]

The issue of public fear of nuclear risks has been dominated by a debate over the rationality-irrationality dimension. The bulk of this research has been done on fear of nuclear power facilities. However, two surveys done in Pennsylvania, and other research,[10] indicate that risks from radioactive waste and nuclear power are not differentiated by many people. Therefore, the results of research on fear of nuclear power will be applied to this discussion.

Robert L. DuPont, a psychiatrist, has obtained research support to enable him to investigate further the nuclear power and radiation "phobia."[11] For him, this fear is irrational and based on three basic elements: people do not perceive that they have personal control over risk factors; the outcome of exposure to the risk can be catastrophic; and the risk is difficult to detect and unfamiliar. DuPont uses these three factors to illustrate why people are not fearful of risks such as automobile use and cigarette smoking but are afraid of exposure to radioactive materials. Other researchers in the areas of risk assessment and sociology present alternative analyses, which argue that public decisionmaking processes regarding radiation hazards are no more irrational than the decision processes used by the technical experts promoting nuclear technology[12] and that there are strong indices of rationality in people's pattern of concern.[13]

Clearly, the rational-irrational debate is part of a labeling process with political ramifications. If public fear is indeed irrational, then more funding should be provided for public education, and public participation in nuclear technology decisions can be discounted as primarily obstructionist. If, however, the fear can be defined as rational, then public participation proponents can more effectively argue for increased public control over the entire process and for the legitimacy of a "go slow" approach.

In fact, in the context of the LLRW issue, the rational-irrational argument, aside from its political ramifications, is basically misplaced. Judgments of rationality hinge on the adequacy of decisions made on the basis of sound, expert knowledge. The problem with low-dosage nuclear materials is that they carry some degree of risk but the magnitude of that risk is largely conjecture and open to challenge:

There appears little consensus among scientists over the health effects of exposure to low-level radiation. In an extensive study done by the Committee on the Biological Effects of Ionizing Radiation issued in 1980, the authors reported that health risk estimates are based on incomplete data and involve a large degree of uncertainty, especially in the low-dose region.[14]

Furthermore, estimating the outcomes of exposure to low levels of radiation is difficult: "The problem is that the numbers of cancers produced per unit dose are so few that they cannot be detected at the levels of exposure to environmental sources."[14] However, there is some evidence that certain genotypes are more susceptible to health problems resulting from exposure to low-dose radiation than are others.[15] In other words, the material is dangerous, the level of danger is essentially conjecture, but some people are more likely to be affected than others, and if you are one of those affected it is likely to be serious.

This degree of ambiguity goes far toward explaining public fear of radiation. It also explains why bringing in more scientific experts will do little to allay public fear: while it cannot be unambiguously proven that low doses of radiation cause specific problems in large populations, it likewise cannot be proven that it does not contribute significantly to health problems. Fear in this case is certainly rational, and it makes little difference that radiation exposure from cigarettes is greater than that from nuclear power plants or that hang gliding involves a higher risk than living near a waste facility. Given the lack of expert consensus, the safest response is to avoid the material if possible.

There are a number of other reasons why public fear of LLRW is substantial. First, nuclear materials tend to be associated with doomsday outcomes. Second, the effectiveness of hazardous waste disposal in general can be questioned. Third, influential entertainment films project an unabashedly antinuclear stance. Finally, the long-term management required for radioactive wastes is viewed by many as unworkable in practice.

Nuclear materials suffer from guilt by association by being cognitively linked to war, "the bomb," and the potential for catastrophic outcomes. For example, in a recent survey done in a Pennsylvania community being considered for a LLRW waste reduction facility, some responses took on the following tone: "We still do not fully understand the effects of the Nuclear Age! People have had questions to ask but NO one to ask them to. If the Russians were to surprise attack, 5% to 10% of the people would be able to react. That's all." Mitchell reports that these kinds of doomsday associations explain a significant amount of variance in attitudes toward nuclear power.[13] Apparently these kinds of fears have been reinforced by the Three Mile Island accident.[10]

Waste of any kind has a bad reputation due to the many instances of waste site problems continually highlighted by the news media. Love Canal, dioxin contamination in Missouri, radioactive slag turning up in building materials, the Superfund "hit list," and other events provide ongoing validation to the public's suspicion about the whole waste handling issue. Sensationalist headlines applied to LLRW, such as "The Deadliest Garbage of All,"[16] not only trigger fears of risky wastes in general but assure the public that their "special" fears of radioactive materials are well founded.

The entertainment media have promoted public fear of nuclear technology, and a distrust of those responsible for that technology, with such films as *The China Syndrome* and *Silkwood*. In the above-mentioned survey, one respondent, in response to an open-ended question on what, if anything, frightened her about LLRW, answered: "Death! See *Silkwood*!"

The radioactive waste issue demands long-term management considerations that are beyond the scope of most people's planning experiences. In two surveys done in Pennsylvania, this single issue was raised by a majority of male respondents as a primary reason why all the technical planning in the world could not insure long-term health and safety. These respondents simply do not trust a technology the soundness of which is predicated on management beyond a single generation. Given high levels of public fear, ongoing reinforcement of that fear by the news and entertainment media, and distrust of long-term waste management, the promotion of public cooperation with siting agencies appears to be a formidable task indeed. However, radioactive wastes carry one further, unusual burden: they have organized interest groups dedicated to eradicating their major generating source, nuclear power plants.

Feeding the Flames: The Role of Antinuclear Activists

The nuclear waste issue has long been defined as the "Achilles' heel" of the nuclear industry. If a satisfactory solution to radioactive waste disposal problems is not forthcoming, the application of nuclear technology will have to be curtailed. Edward Markey, an antinuclear congressman from Massachusetts, intends to use his political power to "finish my efforts to abolish nuclear power in this country."[17] One direction this has taken is attempts by antinuclear politicians and activists to define low-level radioactive waste from power plants differently than that from industry and medicine. The effective separation of these wastes would reduce criticism that the antinukes are threatening social benefits, such as medical treatment and research using radioactive materials. However, research scientists, industry, and regulators counter that this separation would make waste disposal inefficient and costly and do nothing to protect the public from wastes emanating from power plants.[18]

High levels of public fear make it extremely easy for antinuclear activists to help ensure that very few localities seriously consider hosting a LLRW disposal site. Over the past two years I have attended over 20 public discussions of LLRW in Pennsylvania, plus public meetings called in response to a proposed LLRW treatment facility in a small southwestern Pennsylvania community. In many of these meetings, it is clear that the antinuclear message is what the bulk of the public wants to hear, and that it is viewed by the public as an instrument to use to stop siting should that become necessary. In a series of public meetings, one antinuclear scientist argued that radioactive fallout is the cause of falling SAT scores in the U.S. and that infant death rates had risen dramatically after the Three Mile Island nuclear reactor accident. Even though the data he presented did not support his arguments, a majority of those present quickly shouted down any potential detractors.[19]

Even in the rare instance when a community volunteers to host a LLRW disposal site, antinuclear activists can play a role in delaying or stopping the siting process. The economically strapped town of Edgemont, South Dakota, viewed a LLRW disposal site as a source of jobs and revenue. The town had been a uranium mining and milling town, and its familiarity with radioactive materials translated into less fear. Almost 70% of the county and 80% of Edgemont's citizens voted in favor of the proposed disposal facility in June 1984. However, antinuclear activists forced the issue to a statewide referendum, in which a majority of the state's voters favored statewide approval before South Dakota could enter into a nuclear waste compact with another state and approval by a majority of voters before any private industry could be licensed to open a disposal site.[9] In effect, Edgemont's LLRW site has been scuttled.

It is important to stress that antinuclear activists cannot be considered the "cause" of the failure to implement the LLRW Policy Act. Public fear and nuclear industry blunders provide a ready market for the antinuclear perspective. In fact, in many cases, communities faced with the possibility of LLRW invite antinuclear activists to speak.

The Fundamental Issue of Trust

Unquestionably, the basis of all the fear factors and the viability of the antinuclear crusade is the lack of trust of the nuclear industry and its governmental regulators. Substantial data indicate that the Three Mile

Island accident shook public confidence in the nuclear industry.[10,20] Other issues that have damaged the industry's credibility are reports about the Atomic Energy Commission's possible involvement in covering up information about radiation damage to people resulting from the early bomb tests; the Kemeny report and its criticism of the Nuclear Regulatory Commission's handling of the Three Mile Island accident; and public arguments from former nuclear proponents, such as Admiral Hyman Rickover, that nuclear power plants should be abolished.

If the public does not trust the nuclear industry or its regulators, it becomes very difficult to muster support for a project involving radioactive risks. Any attempts by industry or government to promote a given technology, such as a waste facility, will be immediately met with suspicion and even cynicism. In this climate of opinion, information and education programs can do little to foster public cooperation. The ready availability of antinuclear scientists provides an avenue of defense that the public can use to combat information provided by industry and government regulators. Since there is substantial disagreement within the scientific community concerning risk levels, the public is much more likely to listen to those who paint the most pessimistic scenarios.

Summary

Low-level radioactive waste is clearly a "special" problem. The failure of the 1980 Low-Level Radioactive Waste Policy Act should come as no surprise. High levels of public fear, fed by sensationalized newspaper articles and antinuclear entertainment films, a lack of faith in the ability of people to manage anything beyond a generation, the availability of antinuclear activists, and a general distrust of the nuclear industry and its regulators, bode ill for those promoting LLRW disposal siting. There is little reason to believe that the path to the 1992 deadline will be any smoother than that taken toward the 1986 deadline.

In the face of public intransigence concerning LLRW disposal siting, public planners and policy analysts continue to promote better public involvement as the primary solution.[8] However, the elaboration of public involvement programs has not increased public cooperation. The notion that participation leads to cooperation may be based on certain mythical beliefs. An overview of how public participation programs have evolved in response to intransigence dramatically illustrates the dilemma created when democratic ideals are played out in the arena of conflicting private, local, ideological-group, and corporate interests. Public opinion data on the participation issue sheds further light on the problem of eliciting public cooperation in the establishment of LLRW disposal sites.

THE EVOLUTION OF PUBLIC PARTICIPATION PROGRAMS

The Basic-Input Approach

The beginning of the public participation revolution, at least as it concerns environmental issues, is the 1969 National Environmental Policy Act. The basic public input format mandated by that act includes an opportunity for written comment on any government proposal that may have negative local impacts and a provision for public meetings at the proposing agency's discretion. Exactly how this kind of public input is to affect decision making has never been made clear. The program is based on the argument that participation serves the following functions: participation is a fundamental element in realizing democratic ideals; involvement in decision making increases support for the eventual decision; participation can reduce tensions and control conflict; and the public may provide perspectives dealing with crucial issues overlooked by the technical specialists.[21] Reflected in these assumptions is the American liberal faith in communication and interaction as solutions to problems of conflicting interests. Compromise politics is the goal.

However, it has become increasingly obvious that the simple public input process, especially when applied to issues of toxic or radioactive wastes, has not fulfilled the above-stated functions very adequately. The public tends not to interpret participation programs as exercises in democratic decision making, because they do not trust sponsoring agencies or the technical experts representing industry and government regulatory agencies.[22-24]

Further, there is evidence indicating that public participation programs dealing with waste issues may actually result in more negative attitudes than existed before the discussion process.[25,26] Sponsoring agencies apparently have difficulties in effectively communicating need and assuring public safety. The public involvement process can actually heighten citizen awareness of risks, convince them that either the resolve or the technological means to neutralize these risks is not available, and result in well-formed negative attitudes where previously there may have been honest ambivalence or no opinion at all.

Also, rather than abating conflict, public participation programs may enhance hard-line positions. There is some evidence that the existence of public controversy per se increases opposition to a technology, even when proponents and critics are equally vocal.[27] This outcome is especially likely when the technology is poorly understood or when the public perceives a lack of expert consensus about the risks involved. Both of these conditions characterize the LLRW issue. Apparently, when faced with potentially high risk plus ambiguity, the public tends to choose rejection as the safest option.

When the public role is limited to that of providing general input to be used in some relatively unspecified manner, a sense of futility is not an unreasonable response. Rather than approach the participation program as an arena for information sharing and compromise, the public may use the opportunity to exercise the only real power they have, the power to stop the project completely.

Finally, although it is possible that the public's nontechnical viewpoint may provide useful insights, few public participation scholars have noted the problems involved in trying to get meaningful public input on poorly understood technological issues. Aside from voicing general concerns about clean air, water, and soil, along with fears of truck traffic hazards, exactly what can the public contribute to decisions about LLRW? What proportion of the public, even the educated public, understands the basics of soil chemistry, the movement of ionizing radiation through different materials, the relative strengths and weaknesses of containment in certain kinds of clays versus various types of engineered barriers, the virtually insurmountable problem of controlling water infiltration in any kind of disposal design? Even the experts conflict on many of these issues. The most enlightened public must certainly wonder what utility their input has. Again, under the impetus of fear, a lack of trust, and inconsistency of expert opinion, they may be expected to exercise the only real decision making option they have—to say no.

This basic-input public participation format can be viewed as fatally flawed from its inception. In the face of greater public consciousness of environmental problems, and greater political sophistication in general, strategies promoting input without impact have little chance of success. Simple input falls far short of significant democratic participation in decision making. The actual sharing of power is not an aspect of this process. The basic assumption seems to be that the opportunity for input will produce satisfaction and satisfaction will result in cooperation.

In the face of the failure of the basic-input approach, participation scholars, in the late 1970s and early 1980s, began exploring other means to promote public cooperation. The emphasis on equity emerges.

Equity Approaches—From Incentives to Compensation to Benefit Sharing

The use of material incentives to regulate human conduct has a long and rich history. At least three dominant behavioral science traditions— macroeconomics, social exchange theory, and operant behaviorism—are based on the premise that rewards control behavior. Employers have developed varied incentive schemes in attempts to increase employee motivation and productivity. It should come as little surprise that some policy analysts and behavioral scientists have decided that the most effective way to elicit public cooperation on issues involving risky or noxious technologies is to offer some kind of tangible incentives to restore the equity violated when a local community is asked to bear a burden produced elsewhere.[28,29]

The transition from a human relations emphasis to an equity approach has not been abrupt or all-encompassing. Some equity proposals retain a human relations flavor. For example, Howell and Olsen utilize a social exchange framework to argue for a participation program minimizing costs relative to rewards.[30] However, their discussion of costs and rewards focuses primarily on psychic and social costs and rewards. Costs, for Howell and Olsen, include the expenditure of time by citizens, mental anxiety, personal embarrassment, continuing conflict, and input perceived as meaningless. Their suggestions for rewards include the opportunity to influence public decisions, civic pride and community esteem resulting from involvement, and trust generated by program operators. Howell and Olsen do suggest the possibility of providing tangible rewards in the form of payment of public expenses for attending meetings or payments for serving on advisory committees and task forces. However, their approach falls short of a full-blown equity perspective. Howell and Olsen's fundamental assumptions are still those of the basic-input approach. The crucial variable is satisfaction, which is assumed to be linked to cooperation.

An eloquent statement concerning the use of concrete incentives in the case of LLRW siting is presented by Jordan and Melson:

> The actual sociological, psychological, and economic effects of a LLW facility on a community are difficult to quantify. Thus an incentive program should be considered. Compensation of the host locality and state should be considered for actual harm and perceived harm. A comprehensive incentive plan should encompass compensation for actual damage caused by such a facility. It should redress impacts on local property values, taxes, and public services.[31]

This statement clearly establishes the equity theme. Risky technologies bring the potential for various kinds of costs to the community. In addition, the risky facility usually does not provide local compensation to the degree that an ordinary business enterprise can in the way of jobs, gifts, and tax payments.[32] Therefore, a specific plan should be considered that provides economic incentives sufficient to restore equity. Compensation packages make local interests at least as well off as before, while incentive schemes make them better off.[33]

However, it is unlikely that the provision of material incentives will go very far in diminishing public intransigence in the face of LLRW sites. The fundamental issues of insuring public health and safety and of establishing public trust of operators and regulators are not dealt with by such a program. Furthermore, the provision of incentives to affected communities gives the professional opposition ammunition to charge government with trying to buy off the citizenry. Since such incentives will appeal primarily to those who have few resources, government can be further charged with pinning the burden of hazardous wastes on the backs of the disadvantaged. These kinds of accusations have already been leveled at incentive schemes.

Power Sharing Strategies

Neither the basic-input format nor the equity approach to public participation provides real control of decision making to those facing a risky waste facility. If the basic issue is public distrust of the technology and those promoting it, then there is no compelling reason why these techniques should be successful, since they fail to address those issues. However, real community input into site selection, design, operation, and closure decisions has the potential to demystify the process and to foster cooperation. There are basically two power sharing approaches: a weak strategy, proposed by academic policy analysts, and a strong strategy, proposed by somewhat radical environmentalists and community activists.

The work of Morell and Magorian[5] may be the most adequately developed example of a weak power sharing approach. After providing a thorough and insightful analysis of the siting crisis, they advocate a siting strategy that includes negotiations between the developer and the local community, for both site-related inputs and compensation, with the backup of state override in the face of local intransigence. Effective negotiation, according to Morell and Magorian, is contingent upon the power of the local community to compel the developer to "do something." In other words, the local community must have some input into decision making. However, the actual extent of that power sharing is never adequately delineated. Although decrying preemption, the threat of state coercion is retained as a last resort. The Morell and Magorian type of strategy has perhaps been applied most completely in the Minnesota hazardous waste management program. After years of local negotiation, Minnesota is still far from having a site.

The strong power sharing approach is exemplified in the work of Resnikoff[34] and Hurley.[35] Resnikoff passionately argues the case for not trusting the regulating agencies and their employees in cases of radioactive waste siting. For him, the only viable options are to use the vote, the courts, and civil disobedience to insure public safety. Citizen participation is not viewed as effective unless coupled with citizen control over the regulatory agencies plus the provision of funds to local communities to permit their hiring independent experts.

Hurley advocates specific measures that allow some measure of local control. The following are examples of his suggestions: an elected or appointed community member on the decision-making body of the facility, who would be charged with making regular reports to the community; the provision of trained community personnel, who would have access to the facility at any time; mandatory negotiations between the operator and a broad-based community group that has the last say on whether the siting proceeds or not; the provision of funds to communities to allow them to hire their own geologists and health experts; providing the community the power to select that site operator they most trust; and the provision of funds for ongoing health screening.

These proposals, if implemented, would give communities somewhat effective means to protect their own health and safety. The state of Texas has incorporated some of these concepts into their LLRW plan.[36] However, a representative of the Texas Advisory Commission, in a presentation given at the 1986 Waste Management Conference in Tucson, Arizona, indicated that their program is presently stalled due to public and political opposition.

INCENTIVES-COMPENSATION VERSUS POWER SHARING: WHAT DOES THE PUBLIC WANT?

In the heat of speculation about how to handle public intransigence, there has been a noted absence of serious attempts to determine how the public views these issues. In the summer of 1985, a random sample of Pennsylvanians (N = 810, a 57% return rate) completed a questionnaire dealing with policy issues related to LLRW disposal siting. Fifty-two percent of the respondents were males and the rest females. The sample was disproportionately married, over age 30, homeowners, people with a high-school or better education and a family income of $20,000 or more. Thirty-four percent were blue-collar workers; 55% were in managerial, professional, technical, and sales occupational categories. The questionnaire included an introductory letter explaining the important policy issues along with a full front page cover explaining what LLRW is, the

1980 LLRW Policy Act, and the situation in Pennsylvania with regard to these factors. Respondents were asked to make judgments about various incentive-compensation, community involvement, and direct health and safety issues. If equity issues are the major motivating factors driving community opposition, then incentive-compensation issues should be considered most important. If, however, local intransigence is some function of distrust of operators and regulators, then power and control issues should be viewed as crucial.

Five incentive-compensation options were offered:

1. How important is it that property values be protected?
2. How important is it that local agricultural prices be protected?
3. How important is it that tax relief be provided to those living near a LLRW disposal site?
4. How important is it that the site operator formally agree to hire locally and purchase local goods and services?
5. How important is it that the community receive income from a surcharge placed on the waste?

To assess the importance of community power options, the following items were included:

1. How important is it that a community faced with the possibility of siting have input into the decision concerning who actually operates the site?
2. How important is it that a community have input into site construction decisions?
3. How important is it that locals be trained to monitor the site on a regular basis?
4. How important is it that locals be given the power to shut down the site should health and safety problems be encountered?
5. How important is it that the state provide funds to such communities so that they can hire independent experts to check on the technical information provided by the agency responsible for siting?

In addition to the above, two items dealing directly with health and safety issues were included:

1. How important is it that a yearly medical survey of the community be provided to check for health effects related to disposal operations?
2. How important is it that a restricted access road be included that completely bypasses residential areas?

The above questions were asked in two different formats. Respondents were first asked to indicate how important these issues are to them and then they were asked if they thought each of these issues will help promote local cooperation with the siting agency.

Table 1 summarizes the distribution of responses to the importance items and the "will it help" items for the community power, incentive-compensation, and health and safety options. The most noteworthy finding reported in Table 1 is that a large majority of respondents pick the extreme category of importance for every item except the three incentives of local tax relief, agreements to hire and buy locally, and a surcharge on the waste to be returned as revenue to the community. These same items also are chosen significantly less often as definitely helping promote cooperation between the siting agency and the local community.

The items viewed as most important and as most likely to promote cooperation are those dealing with policies that grant control to the community over the site operation process, the specific health items, and one incentive item: guaranteed property values. Only three options are viewed as definitely promoting cooperation: local monitors, the ability of the community to shut down the site in case of problems, and guaranteed property values. It can be argued that these three items reflect a common underlying reality from the point of view of the respondents: if they do not trust the technology, the operators, the regulators, or any combination of those, then site failure can be anticipated. Site failure can be effectively dealt with, from the community's point of view, by (1) detecting the failure as soon as it occurs, (2) shutting down the site as soon as possible, and (3) insuring one's property values in case the failure is dramatic. This pattern of results reflects fear based on a lack of trust.

The nature of this lack of trust can be further specified by answers to the following items: in whom do they have the most confidence to operate the site, whom do they trust to regulate the site, and whom do they trust to represent their interests in dealings with siting agencies?

Three options were given for a potential site operator: a state agency such as the Department of Public Health or the Department of Environmental Resources, a state authority such as the Turnpike Authority, or a properly licensed private industry. Fifty-nine percent of the respondents chose a state agency, 9% chose a state authority, 21% chose private industry, and 11% picked "other." This pattern of responses indicates little confidence in private industry. The preference demonstrated for a state agency probably relates to the respondents' perception that this is an option than can be held more accountable by the citizens.

The item dealing with potential site regulators asked the respondents to express their degree of confidence in six potential regulators: the Nuclear Regulatory Commission (NRC), the U.S. Environmental Protection Agency (EPA), the U.S. Department of Transportation (DOT), the state Department of Environmental Resources (DER), a special state agency created for just this purpose, and trained locals. An explanation of who these agencies are and whether they are federal or state was included. Table 2 presents the distribution of responses.

Two aspects of Table 2 require highlighting. First, NRC, EPA, DER, and a special state agency all elicit high to moderate evaluations by a majority of the public. Second, the public, by a wide margin, prefers

Table 1.   The Distribution of Responses to the "Importance" and the "Will it Help" Questions[a]

Power/Incentive/ Health Items	Importance				Will It Help		
	Extreme	Somewhat	Not-Too	Not	No	Might	Yes
Site operator	84	12	1	2	5	47	47
Construction	65	27	4	3	6	49	47
Local monitors	85	10	2	2	4	38	57
Shut-down power	79	14	3	3	6	35	58
Money to hire experts	76	18	3	2	6	44	49
Disposal method	77	17	2	2	6	49	43
Property value	91	7	1	—	8	40	50
Agricultural price value	79	14	2	2	10	47	40
Local tax relief	43	34	11	9	16	52	30
Buy locally	58	28	9	3	9	50	39
Surcharge	51	34	8	5	10	53	35
Health surveys	84	11	2	1	9	48	42
Special access road	75	17	4	3	8	43	48

[a]In percentages; N = 810. The percentage total does not equal 100 because of missing cases.

Table 2.   Degree of Confidence in Potential Site Regulators[a]

Potential Regulators	Degree of Confidence			
	High	Moderate	Low	None
Nuclear Regulatory Commission	23	32	18	16
Environmental Protection Agency	24	33	18	13
U.S. Department of Transportation	4	18	33	31
Department of Environmental Resources	20	41	19	9
Special state agency	16	36	21	15
Trained locals	37	31	13	8

[a]In percentages; N = 810. The percentage total does not equal 100 because of missing cases.

trained locals as site regulators. Once again, the public's preference for local control is obvious. On the other hand, these data tend to caution against a blanket statement about high levels of public distrust in state and federal agencies. What is clear is that the two options that permit control close to home, the DER and trained locals, are clearly preferred over the others.

Finally, the public was asked whom they most trusted to represent their community in negotiations with those responsible for siting. Six options were offered: county commissioners, a special task force appointed by the governor, local citizens, local officials, a local referendum, and town meetings. Table 3 presents the distribution of responses to that question. The previously noted pattern is reproduced in Table 3; that is, options denoting local control are given higher trust scores than those offering other choices. These options include local citizens, a local referendum, and town meetings. The importance of citizen control is highlighted by the small percentage of respondents who say they have high levels of trust in local officials. Clearly, these citizens want control in their own hands. Constituted authority at any level apparently is suspect. The implications of these data are disturbing: even if siting agencies attempt to negotiate with local authorities, this will fail to impress many local citizens. It is difficult to imagine what kinds of community control would convince locals that their interests are being protected. Referenda provide the opportunity to say no, and any ad hoc local citizens' group can be challenged as being unrepresentative.

These data indicate that the basic issue underlying public opposition to LLRW is trust, not equity. The public apparently primarily trust themselves to protect their own health and safety from this "special" risk. The data also can be interpreted as indicating that the public participation

Table 3.   Distribution of Responses to the Question, "Who Do You Trust to Represent the Community"[a]

Potential Representatives	Level of Trust			
	High	Medium	Low	None
County commissioners	10	31	30	16
Special task force	15	30	22	20
Local citizens	41	27	12	7
Local officials	14	36	23	14
Local referendum	34	25	14	13
Town meetings	35	29	13	9

[a]In percentages; N = 810. The percentage total does not equal 100 because of missing cases.

revolution is alive and progressing vigorously. Options that grant local people substantial control might enhance cooperation with those responsible for siting. However, these results are open to still another interpretation. Granting local communities the right to select the site operator, money to hire their own experts, and the power of referendum almost ensures that a site will never get established. Besides using the referendum as a legal way of saying no, community members can decide they do not trust any operator offered to them, and they can hire experts who discount any information provided by the siting agency.

CONCLUSIONS: PUBLIC PARTICIPATION – A SAFER, MORE EQUITABLE OUTCOME?

The primary goal of citizen participation in decisions involving risky technology is to help ensure public health and safety and to provide community input concerning potentially inequitable policy decisions. Recent innovations in citizen participation promote actual citizen involvement to a degree unanticipated by participation scholars of only a decade ago. However, given the high levels of public fear of radioactive materials, the ambiguity surrounding the technical aspects of exposure to low dosages of radiation, distrust of wastes and waste technology in general, and antinuclear activists ready to capitalize on that fear and distrust, what goals can public participation programs realistically achieve? The most optimistic interpretation of the analysis presented here is that putting real power in the hands of local citizens may result in enhanced cooperation with the siting agency. The most pessimistic interpretation is that communities want greater ability to say no effectively.

It must be noted that even the optimistic interpretation is basically problematic. The degree of local control desired by citizens greatly reduces the autonomy of the waste site operator. In this case, the value of democratic decision making directly encounters the value of private enterprise. This value clash could be circumvented by having the state, instead of private industry, operate the site. The data hint at the possibility that the public might be more comfortable with a state agency as a site operator. However, states have demonstrated little interest in getting involved in the business end of waste disposal siting.

If the existing impasse continues, the government can respond with attempts to force siting through the application of eminent domain. However, this is unlikely given the present climate of opinion concerning both radioactive waste and the value of democratic decision making. A more likely scenario — now unfolding — is that state governments and federal regulators, caught between the vise of the LLRWPAA and negative public opinion, will attempt to develop a technology that will reassure the public that either the hazardous material is safely isolated from the biosphere or, if problems occur, steps can be taken to deal with the problem effectively.

This approach appears to be taking two directions. First, a number of environmental groups, most notably the Sierra Club, are advocating above-ground storage and multiple storage locations.[37,38] Their basic rationale is that the waste should be accessible in case containment problems arise and that no single area should be asked to accept a disproportionate burden. Second, greater emphasis is being placed on engineered containment, such as concrete vaults. At a number of recent professional society meetings concerning LLRW, speakers have pointed to the French underground vault system as a model that may be safer than conventional burial and be more effective in assuring the public that their health is being protected.

The basic problem with an ambiguous, risky technology, such as LLRW, is that there is no way to definitively determine which policies are optimal in protecting public health and safety. Above-ground storage at multiple locations bears its own burdens of proof. Will above-ground structures promote a more cavalier attitude about basic geology and thus increase the probability of environmental contamination? Are such structures vulnerable to storm damage or sabotage? Is worker health and safety compromised? Underground vault construction, even the French model, is susceptible to water infiltration and may increase worker exposure doses. Furthermore, these suggested alternatives are very expensive, especially given that they may not be more effective than conventional shallow land burial, such as that done at Barnwell, South Carolina.

The bottom line for these proposed alternatives is whether they can result in greater public cooperation with siting agencies. This outcome is extremely problematic. Since serious problems are inherent in each approach, it will be very easy to mobilize public fears at any potential site location. Antinuclear activists are not going to stop their crusade simply because of modifications in disposal technology. There is little reason to think that technological innovations will do much to soothe the fears of the public or to significantly reduce their distrust.

None of the options discussed here generates optimism that the LLRW siting impasse will be solved by the new 1992 deadline. In 1980, the federal government tossed the "hot potato" of LLRW to the states, and, in effect, to local communities. However, state and local politicians are understandably even more reluctant to champion LLRW than their federal counterparts. Given the existing climate of opinion on risky wastes of all kinds, the eventual solution to this problem may hinge on reassumption of responsibility by the federal government. The federal government has decades of precedent in establishing public lands to be used for the public good. Geologically superior areas could be designated as federal lands for the disposal of risky wastes. Stringent systems of regulation would have to accompany such a move, and states and local areas would have to have a role in guaranteeing that their interests are protected. The unwillingness of politicians to deal with this issue responsibly may protect local, parochial interests. It does nothing, however, to solve one of the leading public health issues facing modern society. Ongoing efforts to elicit public cooperation must continue to be vigorously pursued. However, it may be time to redirect conventional thinking concerning solutions to the waste siting issue. Local public interests cannot be compromised, but national economic and public health issues must be responsibly addressed.

REFERENCES

1. Bell, D., *The Coming of Post-Industrial Society* (New York: Basic Books, Inc., 1973), pp. 265–266.
2. Nelkin, D., Ed., *Controversy: Politics of Technical Decisions*, second edition (Beverly Hills, CA: Sage Publications, 1984).
3. McKinney, N., "Siting Problem Discussed," *EPA Journal* 5(2): 27 (1979).
4. Bealer, R. C., Martin, K. E., and Crider, D. M., "Sociological Aspects of Siting Facilities for Solid Waste Disposal" (University Park, PA: Department of Agricultural Economics and Rural Sociology [Publication 158], The Pennsylvania State University, 1982).
5. Morell, D., and Magorian, C., *Siting Hazardous Waste Facilities: Local Opposition and the Myth of Preemption* (Cambridge, MA: Ballinger Publishing Co., 1982).
6. "Regional Low-Level Radioactive Waste Disposal Sites: Progress Being Made but New Sites Will Not Be Ready by 1986" (Washington, DC: General Accounting Office [RCED-83-48], 1983).
7. Gettinger, S., "Congress Again Faces Nuclear Waste Crisis," *Congressional Quarterly*, Inc. March 16, 1985, pp. 484–488.
8. Avant, R. V., and Jacobi, L. R., "The Conflict between Public Perceptions and Technical Processes in Site Selection," in Waste Isolation in the U.S.: Technical Programs and Public Participation (Post, R. G., Ed., *Proceedings of the Symposium on Waste Management*, Volume 3 [Tucson, AZ: March 24–28, 1985]).
9. McCaughey, J., "South Dakota Town Dreams of Its Own Nuclear Waste Dump," *The Energy Daily* 13 (17):1, 4 (1985).
10. Freudenburg, W. R., "Waste Not: The Special Impacts of Nuclear Waste Facilities," in *Waste Isolation in the U.S.: Technical Programs and Public Participation* (R. G. Post, Ed., Proceedings of the Symposium on Waste Management, Volume 3 [Tucson, AZ: March 24–28, 1985]).

11. DuPont, R. L., "The Nuclear Power Phobia," *Business Week*, September 7, 1981, pp. 14–16.

12. Slovic, P., Fischhoff, B., and Lichtenstein, S., "Perception and Acceptability of Risk from Energy Systems," in Freudenberg, W. R., and Rosa, E. A., Eds., *Public Reactions to Nuclear Power: Are There Critical Masses?* (Boulder, CO: Westview Press, 1984).

13. Mitchell, R. C., "Rationality and Irrationality in the Public's Perception of Nuclear Power," in Freudenberg and Rosa, eds., *Public Reactions to Nuclear Power* (see above, note 12).

14. Eisenbud, M., "Radioactivity and You," *Environment* 26(10): 633(1984).

15. Committee on the Biological Effects of Ionizing Radiation, *The Effects on Populations of Exposure to Low Levels of Ionizing Radiation: 1980* (Washington, D.C.: National Academy Press, 1980).

16. Stranahan, S. Q., "The Deadliest Garbage of All," *Sci. Dig.*, April 1986, pp. 64–81.

17. "Very Hot Potato," *New England Monthly*, November 1984, pp. 34–39.

18. Brill, D. R., "One Year and Counting," *J. Soc. Nucl. Med.* 26(1):16 (1985).

19. Roth, M., "Fallout from SAT Study Hits Media, Scientists," *Pittsburgh Post Gazette*, November 10, 1984, pp. 1, 4.

20. Walsh, E. J., "Three Mile Island: Meltdown of Democracy?", *The Bulletin of the Atomic Scientists*, March 1983, pp. 57–60.

21. Wengert, N., "Citizen Participation: Practice in Search of a Theory," *Nat. Resource J.* 16(1):23–40 (1976).

22. Abrams, N., and Primack, J., "Helping the Public Decide: The Case of Radioactive Waste Management," *Environment* 22:14–40 (1980).

23. Walsh, E. J., "Resource Mobilization and Citizen Protest Around Three Mile Island," *Soc. Problems* 29:1–21 (1981).

24. Pogell, S., "Government-Initiated Public Participation in Environmental Decisions," *Environ. Comment* 6: 3–15 (1979).

25. Deese, D. A., "A Cross National Perspective on the Politics of Nuclear Waste," in Colglazier, E. W., Jr., Ed., *The Politics of Nuclear Waste* (New York: Pergamon Press, 1982), pp. 63–97.

26. Orr, D. W., "U.S. Energy Policy and the Political Economy of Participation," *J. Politics* 41:1027–1056 (1979).

27. Mazur, A., and Conant, B., "Controversy over a Local Nuclear Waste Repository," *Soc. Studies of Science* 8:235–243 (1978).

28. O'Hare, M., "Not on My Block You Don't: Facility Siting and the Strategic Importance of Compensation," *Pub. Policy* 25(4): 407–458 (1977).

29. Smith, T. P., "A Planner's Guide to Low-Level Radioactive Waste Disposal" (Chicago, IL: American Planning Association [Report #369], 1982).

30. Howell, R., and Olsen, D., "Citizen Participation in Nuclear Waste Repository Siting" (Pullman, WA: Western Rural Development Center, Department of Sociology, Washington State University, 1981).

31. Jordan, J. M., and Melson, L. G., "A Legislator's Guide to Low-Level Radioactive Waste Management" (Denver, CO: National Conference of State Legislatures, 1981).

32. Witzig, W., Dornsife, W. P., and Clemente, F. A., Eds., "Low-Level Radioactive Waste Disposal Siting: A Social and Technical Plan for Pennsylvania" (University Park, PA: Institute for Research on Land and Water Resources, The Pennsylvania State University, 1983).
33. Leistritz, F. L., Halstead, J. M., Chase, R. A., and Murdock, S. H., "Planning for Impact Management: A Systems Perspective," in Murdock, S. H., Leistritz, F. L., and Hamm, R., eds., *Nuclear Waste: Socioeconomic Dimensions of Long Term Storage* (Boulder, CO: Westview Press, 1983), pp. 201-221.
34. Resnikoff, M., "When Does Consultation Become Co-optation? When Does Information Become Propaganda? An Environmental Perspective," in Colglazier, Ed., The Politics of Nuclear Waste (see above, note 25).
35. Hurley, M., "Social and Economic Issues in Siting a Hazardous Waste Facility: Ideas for Communities and Local Assessment Committees" (Fall River, MA: Citizens for Citizens, Inc., 1982).
36. Texas Low-Level Radioactive Waste Disposal Authority, "Siting a Low-Level Radioactive Waste Disposal Facility in Texas: Local Government Participation, Mitigation, Compensation, Incentives, and Operator Standards" (Austin, TX: Texas Advisory Commission on Intergovernmental Relations, 1985).
37. "Low-Level Radioactive Waste," *Sylvanian: Pennsylvania Sierra Club Newsletter*, April/May 1985.
38. Low-Level Radioactive Waste Policy," *Sierra Club*, May 7, 1983.

CHAPTER 11

Low-Level Radioactive Waste:
Can New Disposal Sites Be Found?

E. William Colglazier and Mary R. English

As passed in 1980, the Low-Level Radioactive Waste Policy Act
(LLRWPA)[1] was a law of elegant simplicity. In terse language of only a
few paragraphs, the act specified that each state was responsible for
providing for the availability of capacity to dispose of low-level radioactive waste (LLRW) generated within its borders (except waste generated
as a result of defense activities or federal research and development
activities). States were encouraged to enter into compacts to establish
regional disposal sites. To urge the states to form these agreements, the
LLRWPA provided that as of 1986 any regional compact approved by
Congress could restrict access to its disposal facilities by excluding waste
generated outside the region.

The LLRWPA was atypical of congressional legislation passed during
the last decade in several respects. The law was enacted with relatively
little consideration by Congress of its long-term implications; its passage
was largely an expedient action resulting from pressure generated by the
governors of the three states with operating disposal sites. The act did
not seek a federal government solution to the emerging problem of disposal; responsibility was moved down to the state level. Finally, Congress provided only the broad outlines of policy and resisted trying to
specify the details of its implementation.

The LLRWPA's approach—to encourage a regional system of LLRW
disposal sites based on compacts—seemed economically and politically
sensible. It addressed the fairness issues raised by the governors of South
Carolina, Washington, and Nevada, who contended that their operating
sites should not have to bear the national burden for LLRW disposal
forever, particularly when other regions generated large amounts of

Low-Level Radioactive Waste Regulation: Science, Politics, and Fear, Michael E. Burns, Ed., © 1988 Lewis Publishers,
Inc., Chelsea, Michigan—Printed in USA.

LLRW. The 1986 deadline seemed to create sufficient leverage to make states accept their responsibility, and it was thought that states might be more politically adept than the federal government at securing public acceptance of new disposal sites. The technology did not seem complex, and, with an anticipated six or seven regions, each having a disposal facility, the economies of scale seemed reasonable.

The years since passage of the LLRWPA have not unfolded as its framers expected. To date, not one new site for a LLRW disposal facility is in hand; most states do not even have candidate sites. While seven regional compacts have been negotiated and ratified by Congress, most of the compacts have not yet decided on a host state. A number of states are trying to go it alone or in pairs, rather than collaborate on a regional basis and risk having to accept large volumes of waste from other states. The traditional disposal technology of shallow land burial has been challenged by environmental groups, causing some states to seek a more engineered, but as yet undefined, disposal technology.

By 1985 it had become evident that it would be impossible for the LLRWPA's implied timetable to be met. Both sited and nonsited regions had considerable political leverage — the former in threatening to close or restrict access to existing sites (difficult legally but not practically) and the latter in having many votes in Congress (needed to approve the compacts). Consequently, in December of 1985, Congress amended the LLRWPA to allow seven more years for the states without facilities to develop them, but with milestones and fee surcharges to spur the states into action.

Can regional facilities be successfully developed by 1993? We are pessimistic that the majority of the states and compacts will meet this challenge — in part because of the inherent complexity of siting and licensing, and in part because of the difficulty of overcoming the well-recognized "not in my backyard" (NIMBY) syndrome. Even for states with siting processes that have the necessary ingredients for logistical and political success, these ingredients may not be sufficient. Because of uncertainties in meeting milestones, some generators are considering interim storage plans to manage their LLRW if they should lose access to out-of-state disposal facilities or have substantially reduced access before new sites are ready. If no new LLRW disposal facilities are created, the LLRWPA will be a failed public policy and an enormous waste of time and resources.

In order to understand the types of problems that may emerge in a siting process and to speculate about what is needed for at least some chance of timely success, we will focus on the experiences of three states — Illinois, Texas, and Massachusetts — and one regional compact, the Southeast Compact (including Alabama, Florida, Georgia, Mississippi, North Carolina, South Carolina, Tennessee, and Virginia). All three of the states individually mentioned generate significant amounts of LLRW, particularly Illinois and Massachusetts, as do many of the Southeast Compact states. Illinois and Texas are making progress, especially in that the political leadership of each has determined that it wants a site and has passed an act to that effect, but, as the Texas experience shows, siting a facility can be problematic. Massachusetts is belatedly trying to address its siting problem, but it remains to be seen whether it has the broad-based political will needed to make a serious attempt. The Southeast Compact has developed an approach for selecting a state to host the region's next LLRW facility but, as of mid-1986, the state tentatively identified by the methodology has not accepted this responsibility. Moreover, the most difficult step—that of actually siting a facility—lies ahead, even if the nominated host state proves willing.

ILLINOIS

In response to the LLRWPA, Illinois adopted a comprehensive Low-Level Radioactive Waste Management Act in 1983, specifying a process by which LLRW generated in Illinois will be treated, stored, transported, and disposed of.[2] The act includes requirements for generator registration and annual reports, a fee on LLRW generation ($3/ft³/yr for non-utility generators; $90,000/ reactor/yr for utilities), an interim management plan in case access to out-of-state disposal is cut off, a requirement that a regulatory agreement (Agreement State status) be sought from the U.S. Nuclear Regulatory Commission (NRC); and a detailed siting process. The Illinois governance model is somewhat unusual in that a state agency—the Illinois Department of Nuclear Safety (IDNS)—has been given significant responsibility for both developing and regulating a disposal site.

The generator registration and the draft interim management plan have been completed, the generator fees are being collected, and Agreement State status is being sought. But the siting process is only beginning: IDNS must accomplish a number of complicated and controversial tasks over the next seven years if the state is to have a disposal facility by the new deadline of January 1, 1993. These include developing site evaluation criteria; promulgating rules on contractor selection, waste facility standards, waste treatment standards, waste transportation, and loss compensations; evaluating and selecting from among alternative disposal technologies (traditional shallow land burial is prohibited); selecting a contractor to construct and operate the facility; identifying three sites for characterization and selecting a site; proposing a disposal fee system to the Illinois legislature; and conducting the facility's licensing proceedings.

Accomplishing these tasks will be made somewhat easier by the fact that Illinois, after briefly aligning itself with an eight-state Midwest Compact, dropped out and instead formed the Central Midwest Compact with Kentucky, thereby making it clear that Illinois would be the host state for a new LLRW disposal facility. The Illinois-Kentucky compact came about largely because of pressure from the environmental community in Illinois, which, recognizing that Illinois would probably be a host state in any event, wanted to ensure that the state would have primary control over the disposal site and would avoid large amounts of out-of-state waste. (This pairing between a state with a large LLRW volume and a state with a small LLRW volume, although unanticipated when the LLRWPA was passed, may turn out to be a common arrangement: it is the optimum configuration under which, as part of a compact, a host state can minimize out-of-state LLRW disposal at its site.)

Even though the "host state" issue is resolved in Illinois, as it is not in many states, the siting process still poses many hurdles. Not the least of these is the pervasive — and understandable — resistance of people to having in their backyards an unwanted facility that is perceived as risky to their health and environment. Local communities want to know why the facility is needed, what the risks are, and why they should accept the burden of being the host community for somebody else's waste. Another hurdle, equally pervasive and understandable, is the lack of confidence that private citizens characteristically have in government's ability to be technically competent and to keep its word — a lack of confidence that ranges from mild skepticism to outright suspicion and hostility.

To soften this resistance and to earn this confidence, Illinois intends to do a technically sound job of selecting a site and determining criteria for developing and operating a disposal facility, and to encourage public participation in all steps of the siting process. Although the first cannot be taken for granted, it is, in our view, less problematic than the second. We will thus concentrate on the latter.

The stated purpose of the IDNS's public participation program is "to help IDNS reach wise, just, and fair decisions that can be successfully implemented."[3] The department believes that seeking advice and information from citizens can help both to improve the quality of its decisions and programs and to ascertain what various segments of the public want, and that providing information to citizens on problems and options can facilitate this input. The IDNS also believes that a decision coming from a process that is generally perceived as open and fair is more likely to be accepted by the majority of the public, including many of those who might otherwise oppose the decision. If a significant segment of the public believes its concerns are being ignored, it can virtually stop any project.

To accomplish its goals for public participation, IDNS has decided to:

1. Create a citizens' advisory group on LLRW composed of 15 members representing key constituencies in the state (LLRW generators, environmental groups, the League of Women Voters, universities, etc.). The group, with two neutral facilitators from the Conservation Foundation, will review and comment on IDNS's plans and proposed activities for LLRW management, in order to help point to where problems or consensus may lie.
2. Hold workshops to explore key issues with various groups (representatives of LLRW generators, environmental organizations, etc.).
3. Distribute or provide access to informational materials on LLRW management and the siting process.
4. Circulate draft rules and plans for informal public comment and suggestions, before the rules go through a formal public comment and rule-making process and before plans are adopted.
5. Involve the public—through the advisory group and through public meetings—in the many details to be worked out on the facility siting process (e.g., the criteria for selection of the facility's site, disposal technology, and operating contractor; the terms on which and process by which any impact mitigation and compensation packages will be negotiated with the host community; and the regulations under which facility fees will be determined and any health or property damages claimed will be compensated).
6. Provide funds to the three communities selected as the final candidates to host a LLRW facility, so that they may evaluate the impacts of the proposed facility independently.

The IDNS has begun to implement 1 through 4; 5 and 6 await the initiation of siting. Will siting succeed? Since we worked with IDNS in developing its public participation plan, we are not totally unbiased observers, but we think that it may. The state leadership generally seems committed to developing a LLRW facility; IDNS has a good technical staff, favorable attitudes toward public involvement, and a challenging but feasible framework; and the state has a number of public interest and environmental groups that, while often critical of IDNS and LLRW generators, appear committed to finding a solution to the LLRW problem.

But problems can always crop up. For example:

1. The creation of the advisory group was delayed for a number of months because of political issues concerning its proposed membership and its status as a formal committee; this delay was an irritation to the environ-

mental community in Illinois. And the willingness of the diverse interests on the advisory group to work together remains a fragile commodity.

2. The preemption power of IDNS could possibly be rescinded in the future. A bill giving a designated host community the power to reject a proposed LLRW disposal facility was passed by the state legislature, but was vetoed by Governor Thompson. Although House support sustained the veto, the Senate voted against it.

3. Election politics could affect siting, since the LLRW disposal issue could easily become a liability that no candidate will want to handle — and for good reason, as the Texas experience shows.

TEXAS

Concurrent with the nationwide crisis of 1979, when all disposal sites but the one in Barnwell were closed, Texas was experiencing its own crisis. The Todd Shipyard facility at Galveston, which was the state's largest processing and shipment terminal, attracted public concern because of its growing inventory of LLRW (80% of which came from out of state), its vulnerability to hurricanes and flooding, and an accidental contamination of several Todd workers with radioactive material. Todd first was ordered to reduce its storage of LLRW; then, when it failed to comply, all radioactive material was barred from the site.

Shortly before, in late 1978, Nuclear Sources and Services, Inc. (NSSI) had obtained approval from the Texas Department of Health to build a LLRW processing facility in Houston. However, in the face of vehement opposition from nearby residents and public officials (including several cases of suspected arson and sabotage of company vehicles), the company abandoned the project. In April 1979, NSSI announced that it had bought land about halfway between Dallas and Houston and that it intended to build a storage facility for Texas LLRW. Although the facility would have helped to fill the void about to be left by Todd, local opposition again was fierce. NSSI went ahead and filed an operator's application with the Department of Health, but the application was never acted upon and is now considered to be inactive.

In response to these events and to the LLRWPA, a bill was passed by the Texas legislature and signed into law in June 1981 to create a Texas Low-Level Radioactive Waste Disposal Authority charged with siting, constructing, and managing a disposal facility for Texas's LLRW.[4] In December 1982, the authority contracted with Ebasco Services, Inc., for waste volume projections, a conceptual facility design, and an economic analysis of the facility's life-cycle costs, from siting through decommissioning. In February 1983, the authority contracted with Dames &

Moore to conduct the statutorily required siting study. This study used hydrological, geological, meteorological, and other environmental-impact criteria and a three-phase process that narrowed the location of possible sites to 15 potential siting areas covering 35,000 square miles[2] in all or parts of 105 of the 254 counties of Texas (Phase I); to 280 tracts that were screened to identify 50 potential sites that were then ranked according to environmental and economic/engineering criteria (Phase II); to two sites, one in McMullen County and one in Dimmit County, which were identified on the basis of site-specific data (Phase III).

By January 1985, the authority's board of directors was prepared to name the McMullen County site as the most suitable for a LLRW facility, and, under projections at that time, the facility could have been operational by mid-1988. One month later, after a storm of protest that caused south Texas legislators to introduce a siting moratorium bill and a bill to direct the authority to look for additional sites on state-owned land, the governor met with the board (to which he had just added two new members) and asked that it consider using state land rather than private land for the facility. The authority's staff recommended that the McMullen County site be pursued, but the board unanimously tabled this recommendation.

As directed, the staff then began to evaluate the approximately 3 million acres owned by the state, using basically the same evaluatory criteria and winnowing-out process as before. By June 1985, it had narrowed the field to approximately 66,000 acres, all but 800 acres of which are located in west Texas, and by November 1985 it had identified three possible sites, all in west Texas. According to the current timetable, the candidate site will be identified in mid-1987 and purchased by late 1987. The redirection of the siting efforts will have delayed the scheduled facility opening by about three years.

Was the problem caused by a lack of commitment to public participation? It does not seem so. The act establishing the authority mandates two means of involving citizens who live in the area where a LLRW facility is to be sited: (1) it requires that the authority hold a public hearing after selecting the most suitable site, and (2) it requires that a resident from the area be appointed to the authority's board of directors. It also empowers (although it does not require) the authority (1) to award assistance grants to communities to help reimburse them for costs incurred in evaluating and planning for the facility; and (2) to appoint a mediator to report to the authority's board of directors after holding discussions among interested parties concerning a proposed site. These mandatory and optional methods of involving citizens living in the proposed siting area had not been implemented; the storm of protest rose before this step was reached.

But the authority had used a number of other citizen participation techniques during its two-and-one-half years of site selection: a citizens' advisory committee to counsel the authority's board and staff on LLRW issues, public opinion surveys, public comment periods during the quarterly meetings of the authority's board of directors, public meetings, visits by local officials and private citizens to LLRW disposal sites operating elsewhere in the country, and a dialogue process in two counties.[5] In particular, the authority released public information after every major step in screening, and, as the number of potential sites was narrowed down, it held twenty formal public meetings across the state, all heavily attended. Nevertheless, according to a June 1985 paper presented by Thomas Blackburn, director of special programs at the authority, "The overriding statement of the people is clear: 'We need a site, but not here.' NIMBY ('not in my backyard') has followed us wherever we have gone."[6]

What went wrong? From Blackburn's June 1985 account, it appears that there were several obstacles.

1. The statutory requirement that the authority acquire both the surface and the mineral rights to a large (approximately 300-acre) site, particularly without the power of eminent domain, proved difficult — Blackburn comments that "finding two landowners willing to sell who owned both the surface and mineral estates proved to be a formidable task. Many sites were rejected because of title problems and the inability of the authority to condemn land."[6]

2. Although McMullen County is sparsely populated (0.2 people/mi[2]), a number of landowners and legislators from south Texas with considerable political influence were opposed to the site.

3. In addition, residents of Corpus Christi, although 100 miles away, brought political pressure to bear because they were concerned that the proposed site was only 12 miles from the Choke Canyon Reservoir, which provides drinking water to the city and surrounding areas. (The threat of groundwater and surface water contamination — the threat may be real or imagined — has been a focus for LLRW opposition in Texas and elsewhere: it was a primary concern with Todd Shipyard in Galveston and with the proposed NSSI facility in Houston, and it was one of the reasons for closing three of the nation's six LLRW disposal facilities in the 1970s.[2]

4. The timing for the culmination of the siting process was inopportune. The Texas legislature meets biannually; it had just convened for its 1985 session when the furor over the proposed McMullen County site arose.

5. Siting was also adversely affected by political timing at the national level. Texans had already been concerned that their "go it alone" status might force them to accept out-of-state LLRW shipments once they had an operational facility; the LLRWPA amendments being debated in Congress, which would allow continued national access to existing sites,

increased this concern and also tempered the urgency Texas had felt to site and develop a facility as quickly as possible.

The staff of the authority perceived that there were other, more general problems as well. In Blackburn's account of the failed siting process, he notes that (1) "there is generally someone in the community who has the time, resources, and ability to mobilize the community in opposition to the project . . . "; (2) "politicians tend to stay out of the fray until the last moment because they perceive the issue of low-level radioactive waste disposal as a no-win situation [but] once pressure is brought to bear, they must publicly oppose the siting effort regardless of their personal feelings . . ."; and (3) "the press ranges from supportive to confrontational depending on the circulation proximity to the site . . . [but] tends to exacerbate public opposition by providing a forum for local organizers and politicians . . . [and] it is difficult for the press to differentiate between rhetoric and facts because they usually do not have a technical background and do not have access to technical consultants."[6] Blackburn concludes by remarking, "In spite of our efforts [to respond to the public's concerns], the siting of a low-level radioactive waste facility is heavily influenced by the public and politicians. Their rationale and actions often are not based on the technical merits of the issue and many times override the technical side of the issue."[6]

MASSACHUSETTS

The response of Massachusetts to the LLRWPA began with a 1982 statewide referendum on a bill requiring that no new nuclear power plants or nuclear waste storage or disposal facilities could be constructed or operated in Massachusetts, and no interstate compact for nuclear waste storage or disposal could be entered into by the commonwealth, unless approval was given by majority vote in a statewide referendum and in both houses of the legislature (the General Court).[7] The act, known as Chapter 503, passed by an overwhelming margin. It was sponsored by a number of environmental groups and a western Massachusetts group called the Massachusetts Nuclear Referendum Committee. It is indicative of two prevalent tendencies in Massachusetts: a strong antinuclear bias, and skepticism about administrative competence and the normal channels of representative government.

Before the passage of Chapter 503, the Special Legislative Commission on Low-Level Radioactive Waste had been established by a 1981 act to study LLRW in Massachusetts and make legislative recommendations concerning disposal.[8] (On a separate front, the Massachusetts Executive

Office of Environmental Affairs has talked with other New England states about the possibility of a regional compact, but these discussions have so far been tentative and preliminary.) The commission has 20 members, representing the Massachusetts Senate, the House, the Office of Environmental Affairs, the Department of Public Health, the Massachusetts Municipal Association, the Massachusetts Association of Conservation Commissions, LLRW-generating industries, a nuclear power facility, and the public at large. The commission became staffed and operational in early 1983, and by 1985 it had prepared a draft bill for a Massachusetts Low-Level Radioactive Waste Management Act. The first draft was released in July; a second draft was distributed in September; and, after substantial revisions and considerable streamlining following a series of public hearings held around the state, a final draft was approved by the commission and submitted to the legislature for consideration in 1986.

As proposed to the legislature,[9] the bill establishes a LLRW Management Board composed initially of nine members (seven representatives of public administration, engineering, radiological health, business management, and environmental protection, appointed by the governor for staggered seven-year terms from lists of nominees submitted by relevant organizations; and two representatives of state agencies), and prospectively of two representatives from the area in which a facility is to be sited. The bill gives the board, with its staff, the principal responsibility for the state's LLRW managementwhich, as specified in the bill, includes six phases:

- planning and the promulgation of regulations
- site selection
- operator selection
- facility approval, including review and approval under the Massachusetts Environmental Policy Act (MEPA)[10]
- facility operation and closure
- postclosure institutional control

The bill also specifies that a full-time public participation coordinator shall be part of the management board's staff, and details a number of ways in which public participation must be made possible—in general (e.g., by requiring six geographically dispersed public meetings in the planning phase); with regard to candidate site communities (e.g., by requiring that public meetings be held in possible siting areas and that community supervisory committees be established and given financial assistance when the number of candidate sites has been narrowed to two to five sites); and with regard to the final host community (e.g., by continuing its community supervisory committee and by establishing a citizens' advisory committee to review the proposed project under the required MEPA procedures).

The commission estimates that the first three phases—from initial appropriation to the point at which a complete license application for a disposal facility can be filed—will take approximately four to five years if all goes well. It has not given an estimate of how long the fourth phase—the license review and permitting procedures—will take; that phase is subject to considerable uncertainty. The commission recognizes that it will be difficult to meet the January 1993 deadline for an operational facility and impossible to meet the January 1990 milestone requiring that a complete license application be filed with the NRC, but it has stressed that siting should put environmental safety and public participation before deadlines. Others, such as NELRAD (a consortium of radioactive materials users), support the bill but feel a greater sense of urgency. They caution that, given the possibility of setbacks, the process may take much longer than is anticipated and should be as streamlined as possible.

There are a number of points at which such setbacks could occur. The site selection process includes one stage of six months and another of a year and a half, at both of which, if a majority vote of the management board is not obtained, the stage must be repeated. Also, any aggrieved person may seek review of the board's vote on the facility site through both a special adjudicatory proceeding and judicial review in the Supreme Court of Massachusetts. And the MEPA review could provide extensive opportunity for lawsuits.

On February 25, 1986, the legislature's Joint Committee on Natural Resources and Agriculture held a public hearing on the bill, and of the 26 people who testified, only one—from a radical antinuclear group called Mass Alert—spoke in opposition. Favorable testimony was given by all others, including representatives of environmental groups such as the Environmental Lobby of Massachusetts (ELM), the Massachusetts Nuclear Referendum Committee, the Massachusetts chapter of the Sierra Club, and the Berkshire Natural Resources Council. All of these groups were represented on the commission that developed the bill. There was remarkable unanimity about all of the bill's provisions except the inclusion, in Section 404 of the proposed act, of an explicit requirement that the Chapter 503 test be passed before construction of a disposal facility could be authorized. This provision had not been part of earlier drafts of the bill; its inclusion in this version (and the extensive testimony against it) was done for an ulterior motive—to enable the bill's sponsors to petition the Supreme Judicial Court of Massachusetts for an advisory opinion on the constitutionality of Chapter 503. The Chapter 503 requirement posed a substantial stumbling block to siting, since a proposed facility would have to pass in both houses of the legislature and in a statewide referendum. Some of the bill's sponsors therefore hoped that Chapter 503 would be found unconstitutional, making unnecessary the politically formidable task of having it repealed legislatively or by referendum.

The Supreme Judicial Court's June 1986 opinion found the voter approval process stipulated by Chapter 503 to be unconstitutional under the Massachusetts Constitution as amended, in that the legislature is prohibited from referring any act or resolve taken on its own initiative to the voters for their rejection or approval. It also found the inclusion of the Chapter 503 test in the siting bill to be unconstitutional, in that it encroached on the power of the executive branch in violation of the Massachusetts Declaration of Rights. This opinion was received with relief by a number of the bill's advocates, since it removed a potentially significant obstacle to the act's implementation. Ironically, however, it may present an obstacle to the act's passage. While a number of groups, including environmental groups such as ELM, thought that Chapter 503 imposed an unreasonable and unnecessary impediment to a sound siting process, others, such as the Sierra Club and the Nuclear Referendum Committee, saw it as a reasonable and necessary safeguard in a process that could otherwise become governed by expediency. If the latter's support was contingent upon retaining the Chapter 503 test, then the bill could have a more difficult time getting passed without it.

At least two major hurdles, then, remain before the Massachusetts siting process can be said to be successfully under way: passage of the bill, and a rapid and adequate appropriation to implement it. Governor Dukakis supported the bill when it was introduced but treated it gingerly in the election year of 1986. Although it has the strong backing of a few legislators, so far it has not been seen as a high priority by the legislature as a whole. Even if it passes, initiation of the first phase of siting will await an appropriation, which, particularly given that the act will be expensive to implement, may be slow in coming and may push back the facility siting timetable.

Antinuclear sentiment runs high in Massachusetts and is a popular political gambit there. While nuclear power is the main target, distinctions between nuclear power and radioactive waste often get blurred. Passage of the bill in 1986 was doubtful, and its prognosis thereafter uncertain. Although Chapter 503 — one of the initial responses of Massachusetts to the LLRWPA — has seemingly been put to rest, the attitudes that led to it may continue to confound the state's LLRW facility siting process.

SOUTHEAST INTERSTATE COMPACT

In the Southeast, which generates more LLRW than any other region, eight states—Alabama, Florida, Georgia, Mississippi, North Carolina, South Carolina, Tennessee, and Virginia—each passed legislation agreeing to form a LLRW interstate compact, conditional upon the consent of Congress. On December 19, 1985, consent was granted concurrently with passage of the amendments to the LLRWPA. The state legislation, which is substantially identical to the request to the U.S. Congress for congressional consent (submitted in January 1985 as S. 44),[11] establishes a compact commission composed of two voting members from each party state. It directs the commission:

1. to develop procedures to determine the type and number of regional LLRW disposal facilities needed
2. to provide the party states with reference guidelines for establishing criteria and procedures to evaluate alternative locations for LLRW disposal facilities
3. to develop and adopt procedures and criteria for identifying a party state as a host state to a regional facility
4. if no state volunteers to become the host, to designate the host state by a two-thirds vote.

The legislation provides that the Barnwell facility may be considered the regional facility only until January 1993, and that, at least one year before, another regional facility must be licensed and ready to operate. In developing criteria for the host state selection, the commission is directed to consider "the health, safety, and welfare of the citizens of the party states; the existence of regional facilities within each party state; the minimization of waste transportation; the volumes and types of wastes generated within each party state; and the environmental, economic, and ecological impacts on the air, land, and water resources of the party states."[12] South Carolina will not be asked to serve again until all the others have fulfilled their obligations, as established by the commission, to host a facility.

The commission and its six committees have been meeting at frequent intervals since it was organized in 1983. It has proceeded along several fronts.

1. Information repositories were established within each of the eight state governments.
2. A regional plan was approved by the commission in October 1985. This four-page document establishes the need for an additional LLRW disposal facility for the region; lists the requirements that it must meet

(beyond meeting federal design requirements and annual and total volume requirements, the facility's design would be largely up to the host state); calls for selection of another host state at least seven years before the closing of the proposed facility; and states that the idea of storing LLRW from nuclear power plants at the decommissioned plants had been considered but was rejected, primarily because it would entail a proliferation of waste sites.

3. Dames & Moore was contracted to study the region's LLRW management. Its final technical report was released in April 1986.
4. The host state selection process described below has been implemented.

In April 1985, the commission, at the recommendation of its technical advisory committee, adopted criteria to be used by Dames & Moore in an evaluation of each state's technical appropriateness as the next host state. As subsequently revised by the commission, these criteria include the amount of potentially suitable land in the state, based on geology, hydrology, population, protected areas, and coastal flood plains; the volume of LLRW generated and shipped for disposal in the state by class A, B, or C; transportation distances of LLRW by class A, B, or C; and transportation systems and population densities in the potentially suitable areas. Dames & Moore was asked to arrive at numerical indicators of each state's ability to meet each of these criteria, but was asked not to reveal these numbers until the commission had agreed on factors to weight the criteria. The weighting factors were then applied to Dames & Moore's numbers to arrive at the ranking of each state's technical appropriateness to host the next regional facility. According to the results, which were adopted by the commission in April 1986, North Carolina was ranked first, followed by Alabama, Virginia, and Georgia. The commission had undertaken two parallel efforts to facilitate selection of the host state: (1) it had asked each state to prepare a plan specifying what conditions it would seek (including, e.g., fee surcharges and other forms of compensation) if it were the host state; and (2) it had asked each state to consider whether, when, and with what stipulations it would volunteer to host a regional facility. Although the states were not committed to the provisions of their plans, the commission decided by a close vote that submission of the plans should be mandatory, in order to force the states to think about their bargaining positions. The plans were submitted in January 1986. No state has volunteered to be the host state.

In late May 1986, the commission held a public hearing in each of the four top-ranked states in order to receive testimony on the technical findings and on other factors that should be taken into account in selecting a host state. A number of technical objections were raised at the North Carolina hearing, and in response the commission agreed to have Dames & Moore review their application of some of the criteria. As a result, the commission's vote to select a host state, which had been scheduled for July 1986, has been postponed. North Carolina is irate that only some of its objections are being met. Alabama, Virginia, and Georgia are irate that North Carolina is raising objections to a process that had been agreed upon. The compact remains intact for now—none of the party states has pulled out yet—but the workability of its process is in question.

If a state accepts host status, it will largely be up to that state to determine what type of facility it will have and where and how that facility will be sited. The commission does direct in its regional plan that "public participation at all levels should be solicited."[13] In addition, it has used public participation extensively in its own proceedings, has supplied public participation goals and strategies to the states, and has developed a siting primer. But in the final analysis, the facility's siting process will necessarily be determined by the host state—and as the experiences of Texas, Illinois, and Massachusetts indicate, this is likely to be a difficult task.

PROBLEMS AND PROSPECTS

As these four situations illustrate, the LLRWPA has encountered problems that were unforeseen in 1980, when the act was passed. These problems have appeared at all stages in the envisioned transition to a rational system of regional disposal facilities. They have occurred at the compact formation stage, where regional alignments were delayed by denial ("The 1986 deadline isn't real") and gamesmanship ("If we hold out, maybe we'll get a better deal"). They can be anticipated at the stage of host state selection, where the only states that so far have accepted host state status are those which for the time being are (unwilling) host states to the nation (Nevada, South Carolina, and Washington); those which are going it alone (e.g., Texas); and those that, by joining with states generating only minor LLRW volumes, have similarly opted to have both responsibility for and control over their LLRW disposal facilities (e.g., Illinois and Pennsylvania). And problems are beginning to occur at the site selection stage, as the states that are assuming siting responsibility turn to their communities and try to find technically suitable and politically viable locations for LLRW disposal facilities.

The 1985 amendments to the LLRWPA attempt to address these problems by a combination of penalties and constraints exacted in exchange for a seven-year extension of the original January 1986 deadline. Whether these measures will spur state action (and whether they could have if they had been adopted in 1980 with the original LLRWPA, or whether another political crisis was necessary) is anybody's guess. While the 1985 amendments have succeeded in catalyzing the formation of compacts and may contribute some urgency to selection of a host state, it will still be the responsibility of individual states to resolve the problems of actually siting a facility.[14]

Siting problems in LLRW disposal come from a number of sources, many of which have been alluded to in the case studies given above. They come from a fear of radioactivity, knowledge of past environmental problems, and a mistrust of government and industry. For some individuals, there is hostility to anything having to do with nuclear power. There may also be a blurring in many people's minds of the distinctions between low-level radioactive waste, high-level radioactive waste, and toxic chemical waste. As a result, from publicized problems in handling one (Love Canal, Times Beach), potential problems in handling the others are inferred. The West Valley, Sheffield, and Maxey Flats sites also gave a strong impression to the environmental community that shallow land burial is not an acceptable approach to LLRW disposal, yet it is the approach referenced in the NRC's 10 CFR 61 regulations. While 10 CFR 61 does not preclude the additional use of engineered barriers, environmental groups are suspicious that the cheapest solution will be sought unless special requirements are enacted into law. More generally, there has been a movement toward community control during the past decade that has created an atmosphere of hostility toward all government, and especially the higher levels of government. And the climate of urgency at the national and state levels to site LLRW facilities — while needed to get the states moving — may well exacerbate rather than extinguish local hostility and suspicion.

The states, then, have a difficult task in convincing local communities to host LLRW disposal sites. The LLRWPA clearly maintains that each state is responsible for ensuring that it either has a disposal site or has access to one through a regional compact. But local communities ask why they should become a dump for everyone else's radioactive waste. Can the states successfully overcome local opposition? Perhaps, but not by preemption.

THE MYTH OF PREEMPTION

Within the context of LLRW management, "preemptive power" refers to the legal power of a state agency or authority to override a lower level of government, such as a county or municipality, in making decisions about siting and operating a LLRW disposal facility. But the legal concept of preemption is also a political concept, and, in the latter sense, it can be a myth. When a higher authority attempts to exercise its will over a community in the face of implacable local opposition, this attempt almost invariably fails, whether or not it is backed by a legally valid preemptive power. Robert Avant, in his former capacity as assistant general manager of the Texas LLRW Disposal Authority, and Lawrence Jacobi, general manager of the authority, have identified seven stages of opposition to a LLRW disposal facility siting process:

1. Individual contacts and letters appear when the general public first learns that their area is under construction.
2. Formal petitions and resolutions oppose the activity.
3. Special interest groups form, e.g., South Texans Against Nuclear Dumping. Also, outside groups who are sympathetic to the local cause may join in the opposition.
4. An attempt is made to halt siting progress by legal action—injunctions, restraining orders, lawsuits, etc.
5. Political intimidation or formal political action occurs. For example, political threats may be made against the future of the operating entity, or more formal political actions such as executive orders or legislative actions may be taken.
6. Formal intervention in the licensing process occurs.
7. Public disobedience breaks out.[15] Avant has commented that so far the authority has experienced the first five stages, and, as his coworker Thomas Blackburn has noted, "the public has been extremely effective in opposing our efforts by following these steps."[6]

Texas is an unusual state, but it's not unique in this regard. A letter in *The Christian Science Monitor* for October 17, 1985, commenting on the co-optation of "mainstream" environmental groups, states the following:

The new leaders of the environmental movement are the homemakers, farmers, blue-collar workers, and small business operators who have formed hundreds of grass-roots groups to clean up toxic dumps, end poisonous plant emissions, and block poorly designed hazardous waste sites . . . These groups have not given up confrontation for the genteel art of negotiation but rather have rediscovered the classic forms of protest long abandoned by mainstream environmentalists. No new hazardous waste site has been built in years, due to grass-roots lawsuits. Further, it's because of this new, grass-roots environmental movement that we'll have a stronger Superfund law, improved EPA action, or both. Thousands of average people are learning, in the fight for their homes and families, a lesson that the traditional environmentalists have lost: social progress seldom comes without a fight.[16]

Although the letter's author refers mainly to hazardous waste sites, he could easily be speaking of LLRW disposal sites. His rancor may be extreme, but it is not atypical. However, he fails to recognize another potential problem: while public action is being blocked, potentially more dangerous private action (moonlight dumping of hazardous chemicals, inadequate decentralized storage and disposal procedures for LLRW) may be going on.

CONCLUSIONS

We have no easy answers to the states' siting dilemma, and the suggestions we do have pale before the problem. We put these forward not as a formula for a successful siting process, but as what we think are necessary conditions. It remains to be seen if they will be sufficient.

In designing the siting process, one must keep in mind five basic questions that have to be satisfactorily answered at the state and local levels:

1. Does the decision process sufficiently allow for the views of key stakeholders to be heard?
2. Is the facility truly needed?
3. Is the site selection process sufficiently objective and scientifically credible?
4. Is the facility reasonably safe?
5. Can incentives be provided that would adequately compensate the host community for perceived negative impacts?

All stakeholders must be convinced that their views are being heard and addressed in siting. The determination of need must be convincing to both the state leadership and the local community, in terms both of necessary products and services provided by the waste generators and of a demonstrated commitment to reduce as much as possible the amount of waste that must be disposed of. The site selection process must generally be based on scientific and technical criteria rather than politics; political considerations can play a justifiable role in the site selection if the community is favorably inclined and the site is technically adequate. The question of reasonable safety or acceptable risk must be answered in a way that is credible to the local community. There must also be a demonstrated commitment to the best available technology, a high level of quality control, and rigorous regulatory oversight. If need and safety can be answered, then incentives to entice the community to accept the perceived burden of hosting others' waste disposal must be candidly addressed.

With these five underlying considerations in mind, we view the necessary elements of a siting process for a LLRW disposal facility to be as follows:

1. The political leadership of the state (the governor and the legislature) must have concluded that the facility is truly needed, must have assigned the responsibility for it to a capable state agency, and must be willing to support the agency's efforts. The agency, however, must understand that the siting effort can pose a predicament for an elected politician and must be willing to be sensitive to political considerations and to insulate the governor from some of the political repercussions of siting.

2. The project's technical execution must appear to be, and be, impeccably competent. The staff must be up to the job, and outside peer review must be sought.

3. Although formal decisions should reside with the agency, it must adopt a credible public participation process that goes beyond minimal legal requirements. The power of preemption is an important asset, since it cloaks the agency with greater authority, but, as noted above, in practice the agency will probably be unable to use this power effectively if the community is unalterably opposed.

4. Certain broad issues of LLRW management must be addressed, using some sort of consensus-building process, before the siting effort focuses on candidate sites. The issues to be addressed include source and volume reduction; alternative technologies for storage, treatment, and disposal; and siting criteria. A management plan developed by the agency with significant public participation is an effective means of addressing these issues. A small policy dialogue group representing the key interests and stakeholders can help to point to where consensus may lie on certain controversial issues before the broad public participation effort is undertaken.

5. The two-part purpose of a broad public participation effort is (a) to receive citizen input in order to improve the agency's decisions and discern what the public wants, and (b) to increase the likelihood of public acceptance by having a process that is recognized as open and fair. To achieve this dual purpose, four points about communicating with the general public should be noted:

   a. Public information about siting should be early and widespread; should use different media, understandable language, and timely presentation in manageable amounts; and should come from as many credible sources as possible.
   b. Key people should be involved at key points in the process — formally, through hearings and negotiations, but especially informally, through meetings, workshops, dialogue groups, and phone conversations. These key people will vary, depending upon the point in the process.
   c. The process should not promise a great deal and then fail to deliver. It should be realistic about funding, staffing, workloads, and timeliness and also about how open it will be.

d. Unknowns should be acknowledged; bluffs do not work for long and lead to suspicion and hostility.

6. Funds should be given to candidate host communities to enable them to evaluate independently the proposed project with their own consultants and to formulate their own requirements for making the facility "acceptably safe." A socially acceptable technology may be more rigorous than what technocrats deem necessary. The community must be asked what it thinks is necessary, in terms of technology design, monitoring, assumption of liability, and local control, to make the facility an acceptable neighbor. (Recent studies have indicated that local control, such as through an oversight committee that could shut down operations for certain reasons, is more important to the public than incentives in making a facility acceptable.)[17,18]

7. An impact mitigation and compensation package should be negotiated with representatives of the host community when the community as a whole is comfortable with the issue of safety. (This, of course, poses the thorny questions of which representatives can adequately speak for the rest of the community and who should select those representatives. We do not have any answers to these questions; we can only comment that in keeping with the community's autonomy, the selection should be done at the local level, and that to avoid charges of trying to buy the community off, negotiations on a mitigation and incentives package should be begun *after* safety issues have been resolved.)

8. The inherently unequal distribution of the outcomes of siting should be acknowledged.

The last point may be the most difficult of all, and it suggests that siting a facility which no one wants "in my backyard" may be one of the most difficult tasks in a democracy such as ours. John Rawls, in the first chapter of *A Theory of Justice*, comments that "justice is the first virtue of social institutions,"[19] but he goes on to note that different political systems can have different conceptions of justice. In the United States, our system is built on the conception of justice as fairness. We do not like to think we are taking on — or giving to someone else — more than a fair share, especially when what's being shared is perceived to be risk, not benefit. But the LLRW disposal process leads ineluctably to an unequal distribution of perceived risk: many people benefit (through electricity, medical treatment, smoke alarms, and other technologies) from one process or another that involves LLRW, but only a handful of communities will end up with a LLRW disposal facility as a neighbor.

The progressive pieces of legislation on this issue reflect an increasing uneasiness with maintaining equity. What starts as simple fairness (each state should take responsibility for its own LLRW, as the two-page

LLRWPA of 1980 holds) proceeds to a more complex kind of fairness (any party state that hosts a regional LLRW disposal facility shall not be designated as host for an additional facility until each party state has fulfilled its obligation to have a regional facility operated within its borders, as the 24-page Southeast Compact holds) and then evolves to minutely detailed attempts to maintain procedural fairness in the face of an inevitably unequal distribution of perceived risk (the Massachusetts LLRW bill is 49 pages long). Uneasy as we Americans are with the notion of a process that has unequal results, even if the process is not in itself inequitable, we are even more uneasy with what may appear to be a thinly veiled utilitarian conception of justice: the greatest good for the greatest number. Public decisions entailing the sacrifice of individual interests to broader interests are not unprecedented in this country; in fact, they are widespread (witness our tax system). But when a few have been singled out to make sacrifices, these sacrifices largely have been confined to volunteers (workers who knowingly assume high-risk jobs, sometimes with special compensation), to wartime (drafts by lottery and assignments by platoon), or to public works projects (displacement with compensation when a river is dammed or a city block is subjected to "urban renewal"). Such public decisions can be, and often are, inequitable as to their outcomes. (The Quabbin Reservoir in rural western Massachusetts was created out of farmland to supply water to urban eastern Massachusetts.) But these outcomes generally are recognized costs or hazards, not risks of arguable proportion; they generally involve money or real property, not health; and they usually affect the current population only, not future generations as well. Many citizens perceive that none of these three characteristics holds with a LLRW disposal facility.

The equity issue, then, turns in part on the perceived risk of a LLRW disposal facility — and perceptions on this vary enormously. The radioactive waste industry and others, such as LLRW generators and state officials charged with managing LLRW, sometimes speak of LLRW as if it were mainly contaminated booties, disposed of easily with no adverse consequences; the environmental community and others in the public at large sometimes speak of LLRW as if it were an unprecedented hazard that requires constant vigilance. And because of this dichotomy, objections by the public to having LLRW disposal facilities nearby tend to arouse negative reactions. The scientists and engineers say, "The public is ignorant"; waste generators say, "The public is ungrateful"; state officials say, "The public is unreasonable" . . . while the public says, "Why me?" Although the technical experts may be confident that the real risks are indeed small, the potential host community and the state as a whole must be convinced that this is the case,[20] and they must also be convinced that they are not being unfairly imposed upon.[21]

The most difficult challenge is to come to a more widely held sense, both in the potential host community and in our broader social and political community, of what actual risks are and are not involved in a LLRW disposal facility, and of how the burden for hosting undesirable but necessary land uses can be fairly compensated. To fail to answer both questions satisfactorily means almost certain failure of siting.

ACKNOWLEDGMENTS

A number of people generously made themselves available to the authors during the preparation of this paper. In particular, the authors gratefully acknowledge the contributions of Teresa Adams, Terry Lash, Diane Mayer, Judy Shope, Richard Smith, Janis Stelluto, and Kathryn Visocki, all of whom provided extensive background information; Robert Avant, John Stucker, Richard Walker, and Susan Wiltshire, who reviewed an earlier draft of the paper; and Mary Redfearn and Betty Moss, who saw the manuscript through several rounds of revisions. The authors are responsible for any errors of fact or interpretation.

Preparation of the paper was supported by the Waste Management Research and Education Institute, a state-sponsored "center of excellence" at the University of Tennessee, Knoxville. A somewhat longer version appeared in the *Tennessee Law Review*, Volume 53, Number 3.

NOTES AND REFERENCES

1. Low-Level Radioactive Waste Policy Act, P.L. 96–573 (42 U.S.C. Sections 2021b–2021d).
2. Illinois Low-Level Radioactive Waste Management Act, P.A. 83–0991.
3. "Public Participation Plan on Low-Level Radioactive Waste Management in Illinois," Illinois Department of Nuclear Safety (November 1985), p. 2.
4. Texas Low-Level Radioactive Waste Disposal Authority Act, Chapter 273, Acts of 1981, as amended by Chapter 692, Acts of 1985.
5. "Siting a Low-Level Radioactive Waste Disposal Facility in Texas: Local Government Participation, Mitigation, Compensation, and Operator Standards," Texas Advisory Commission on Intergovernmental Relations, staff report (n.d.), pp. 11–12.
6. Blackburn, T. W., III, "Low-Level Radioactive Waste Activities in Texas," paper presented at the Radioactive Exchange Decisionmakers' Forum, Charleston, SC, June 57, 1985, n.p.
7. Nuclear Power and Waste Disposal Voter Approval and Legislative Certification Act, Chapter 503, Acts of 1982.

8. An Act Providing for an Investigation and Study by a Special Commission Relative to Low-Level Radioactive Waste, Chapter 738, Acts of 1981.

9. Massachusetts Low-Level Radioactive Waste Management Act, submitted as H.B. 5000 in January 1986 (now referred to as Senate No. 1763).

10. Massachusetts Environmental Policy Act (M.G.L.Ann., Chapter 30, Sections 61–62H).

11. S. 44, Ninety-ninth Congress, First Session.

12. S. 44, Article 4, Section E.

13. "A Plan for the Safe Disposal of Low-Level Radioactive Wastes in the Southeast Region," Southeast Interstate Low-Level Radioactive Waste Management Compact Commission (October 1985), p. 2.

14. This, of course, poses a different kind of problem for the other states in a regional compact: their ability to meet the prescribed milestones will depend largely on the ability of the compact's host state to successfully carry out a siting process, but the other states will have no direct political means to ensure that success.

15. Avant, R. V., Jr., and L. R. Jacobi, Jr., "The Conflict between Public Perceptions and Technical Processes in Site Selection," paper presented at Waste Management '85, Tucson, AZ, March 37, 1985, n.p.

16. Collette, W., letter to *The Christian Science Monitor*, October 17, 1985.

17. Bord, R., "Opinions of Pennsylvanians on Policy Issues Related to Low-Level Radioactive Waste Disposal" (University Park, PA: The Pennsylvania State University Institute for Research on Land and Water Resources, September 1985).

18. Elliott, M. L. P., "Coping with Conflicting Perceptions of Risk in Hazardous Waste Facility Siting Disputes," Ph.D. dissertation, Massachusetts Institute of Technology, Cambridge, MA (1984).

19. Rawls, J., *A Theory of Justice* (Cambridge, MA: Belknap Press of Harvard University Press, 1971), p. 3.

20. It is interesting to note that strong opposition to a radioactive waste facility does not always occur within the proposed host community with support occurring at the broader, statewide level. Instead, the situation may be reversed. This has been demonstrated recently in Oak Ridge, TN, where a U.S. Department of Energy proposal to establish there a facility for temporarily storing and repackaging spent nuclear fuel (a Monitored Retrievable Storage facility, or MRS) has been supported by the local government provided certain conditions are met, but has been opposed at the state level. It has also been demonstrated in Fall River County, South Dakota, where a plan to site a LLRW disposal facility there was approved by a majority of county residents voting in a statewide referendum on the question on November 12, 1985, but was rejected by 83% of those voting in the state as a whole. These two cases may be unusual in that both communities have a longstanding familiarity with radioactive materials (Oak Ridge because of nuclear research and weapons production; Fall River County because of uranium mining). However, they also suggest that, in at least some cases, impact mitigation and compensation packages offered to host communi-

ties, together with related economic benefits, may help to quell the reservations of those who stand to be the primary recipients of those benefits, but not the reservations of those at the broader, statewide level.

21. Washington and Oregon offer an example of how responsibility for undesirable but necessary facilities can be distributed: apparently, these two states have at least a tacit agreement that Washington will continue to host a LLRW disposal facility for the region and Oregon will continue to host a hazardous waste facility.

CHAPTER 12

Biologic Effects of
Low-Level Radiation

Rosalyn S. Yalow

INTRODUCTION

In a 1981 report to the Congress of the United States by the comptroller general, "Problems in Assessing the Cancer Risks of Low-Level Ionizing Radiation Exposure,"[1] it was pointed out that since 1902, when cancer was first attributed to overexposure to X-rays, the U.S. government has spent close to $2 billion (approximately $80 million per year in recent years) on research on the health effects of ionizing radiation. At least 80,000 scientific papers on the subject have been published worldwide. The report states that while much has been learned about the carcinogenic effects of high doses of radiation exposure, scientists are still uncertain how ionizing radiation causes cancer, and how to predict the effects of exposure to low doses.

It should be appreciated that, before World War II, radiation protection was based upon the concept of a "tolerance dose"—a dose below which there were believed to be no harmful effects of radiation. After World War II, bodies concerned with radiation protection generally adopted the "no threshold" hypothesis, and established protection guides and regulations based on the supposedly conservative assumption of a linear dose-response curve. The basis for the changeover was philosophical rather than scientific, in that it was not based on epidemiologic or experimental data that reliably demonstrated that there was increased carcinogenesis at low doses delivered at low dose rates. Rather, it was a consequence of the development of highly sensitive radiation detection equipment and the establishment of health physics programs in the Manhattan Project during World War II which made practical the establishment of guidelines that would not have been

Low-Level Radioactive Waste Regulation: Science, Politics, and Fear, Michael E. Burns, Ed., © 1988 Lewis Publishers, Inc., Chelsea, Michigan—Printed in USA.

possible when exposures were only roughly evaluated, as had been done previously, on the basis of skin erythema doses. Furthermore, the possible exposure of large groups of people to ionizing radiation because of the development of nuclear weaponry, the introduction of nuclear medicine, the increased use of diagnostic and therapeutic radiation, and the anticipated progress in the use of nuclear power provided the incentive to establish tighter guidelines than had earlier been used.

This review will consider some of what is known about health effects and, in particular, the carcinogenic effects associated with low doses of lightly ionizing radiation. In general, studies to be reported include those concerned with possible incidence of malignancies due to: natural background radiation in different regions of the world as well as to elevated radon levels in the home; radiation to patients and staff associated with medical diagnosis and therapy; and radiation exposure associated with nuclear weapons testing. The potential usefulness of animal studies in predicting human radiation-induced carcinogenesis will be addressed. In addition, some consideration will be given to the potential genetic and mutagenic effects of low-level radiation. Since there are tens of thousands of papers in this field, there obviously has had to be considerable selection in the choice of studies to be described.

RADIATION UNITS

Commonly used units for radiation exposure are the rad or the rem. A rad is a unit of absorbed dose or energy absorbed per unit mass from ionizing radiation and corresponds to 0.01 joules/kg. Densely ionizing radiation, such as that associated with alpha particles, protons, or fast neutrons, is more effective in producing deleterious biologic effects than is the lightly ionizing radiation associated with beta, gamma, or X-radiation. A rem is a unit that takes into account the relative biologic effectiveness (RBE) of lightly (low linear energy transfer, LET) and densely (high LET) ionizing radiation. A rem is an absorbed dose that produces the same biologic effect as 1 rad of low LET radiation. Therefore, the units rad and rem are generally used interchangeably for low LET radiation. The RBE is not a true constant for any ionizing particle, even those of low LET, but depends to some extent both on its energy and the biologic effect under observation. For instance, the RBE for alpha-particle radiation is usually considered to be in the range of 10 to 20.

NATURAL BACKGROUND RADIATION

It should be appreciated that radiation in the environment from natural sources is the major source of radiation exposure to the inhabitants of our planet.[2] Our body contains about 0.1 μCi of K-40, a primordial radionuclide with a half-life of 1.3 billion years. It contributes about one-third to this natural background. Another third is contributed by cosmic radiation and the other third by the natural radioactivity of the soil and building materials. The average natural background radiation dose in the United States is considered to be about 0.1 rem per year.

It would seem most reasonable that if we want to determine what effects of radiation are at dose rates and doses comparable to those attributable to natural background radiation, populations should be examined who live in regions of the world where background radiation is higher than that to which most of the rest of the world is exposed. Such a study was performed in China by examining 150,000 Han peasants with essentially the same genetic background and same lifestyle.[3] Half of the group lived in a region where they received an almost threefold higher radiation exposure because of radioactive soil. More than 90% of the progenitors of the more highly exposed group had lived in the same region for more than six generations. The investigation included determination of radiation level by direct dosimetry and evaluation of a number of possible radiation-related health effects, including chromosomal aberrations of peripheral lymphocytes, frequencies of hereditary diseases and deformities, frequency of malignancies, growth and development of children, and status of spontaneous abortions. This study failed to find any discernible difference between the inhabitants of the two areas. The authors of this study concluded that either there may be a practical threshold for radiation effects or any effect is so small that the cumulative radiation exposure to three times the usual natural background resulted in no measurable harm in this population after six or more successive generations. Even if the linear extrapolation were valid, one would not have expected to have detected harmful effects, since the size of the exposed group was comparable to that of the Hiroshima-Nagasaki survivors, and their cumulative exposures over 10 years were only about 10% of the acute exposure of the Japanese survivors. It should be appreciated that among the 82,000 atom bomb survivors, whose mean acute whole body dose was about 25 rem, the incidence of malignancies through 1978 was less than 6% greater than would have occurred without the radiation exposure—4500 cancer deaths would have been expected without additional radiation exposure, and an additional 250 cancer deaths, 90 of which were leukemia, were estimated to be a consequence of such exposure.[4] However, if, as some suggest, the linear-extrapolation hypothesis underestimates effects at low doses and dose rates, then the Chinese study[3] might have discerned such effects.

In different regions of the United States, there are variations in natural background radiation: those residing in the Rocky Mountain states receive on the average an additional ~ 0.1 rem/yr compared to the rest of the population. Frigerio and Stowe[5] have observed that the cancer rates in the seven states with the highest background radiation levels are lower than the average United States rates (Table 1). These data might suggest a protective effect of this excess radiation delivered at low dose rates, although other factors to be considered include decreased oxygen tension or the presence of agents in the soil, water, or vegetation of these mountains that are capable of inhibiting the action of other carcinogenic factors in the environment. Nonetheless, had the cancer incidence or mortality been greater in the Rocky Mountain states, radiation effects, rather than other environmental factors, would have been unequivocally declared by some to be the causative agent. In the Rocky Mountain states, cumulative excess exposure averages about 1 rem for each decade of residence. Thus, even in a group as large as 5 million receiving this excess radiation exposure, genetic and/or lifestyle factors are of such overwhelming causative importance that one cannot attribute variations of cancer incidence or mortality either to advantageous or to deleterious effects of low dose/low dose rate radiation. There are regions of the world, in India and in Brazil, where natural background radiation is up to tenfold higher than usual (~ 1 rem/yr) and deleterious health effects have been looked for and not found.[6-10] It should be appreciated that over 25 years, these exposures equal the acute exposures of the Hiroshima-Nagasaki survivors.

ELEVATED RADON LEVELS IN THE HOME

Recently the demonstration that in homes built along a natural uranium-laden geologic rock formation underlying a number of counties in Penn-

Table 1.   Cancer Rates in U.S. White Population in Rocky Mountain States Compared with the U.S. Average, 1950–1967[a]

	7 Highest States	U.S. Average
Background mrad/yr	210	130
All malignancies[b]	126	150
Leukemia	7.0	7.1
Breast cancer	21.5	25.3
Lung cancer	14.5	20.4
Thyroid cancer	0.055	0.057
Malignancies, age 0–9	8.1	8.5

[a]Source: N.A. Frigerio and R.S. Stowe.[5]
[b]Rates are per 100,000 population per year.

sylvania, New Jersey, and New York, the Reading Prong, indoor radon concentrations may be comparable to or even greater than those found in uranium mines has generated considerable concern. Furthermore, sampling of other houses in New Jersey, not in the Reading Prong, has also revealed elevated radon concentrations. The National Council on Radiation Protection has estimated that, nationwide, 9000 deaths annually from lung cancer might be due to radon in homes. The federal Environmental Protection Agency has estimated the radon-related lung cancer deaths to be in the range of 6000–20,000 per year, quite comparable to the Centers for Disease Control estimates of 5000–30,000 radon-related deaths. These numbers are obtained using lung cancer risk estimates given in the BEIR III Report,[11] which were based on studies of uranium miners. According to the American Cancer Society, it is estimated that there will be 89,000 male and 41,000 female lung cancer deaths in 1986.[12] Let us consider if it is likely that up to one-quarter of these lung cancer deaths could be due to indoor radon levels.

The age-adjusted lung cancer death rate in 1981 for males was 72 per 100,000 population; the corresponding rate for females was 21. In 1930, when the incidence of smoking was much less than in the following decades, the male and female age-adjusted cancer death rates were 4 and 2 per 100,000 respectively.[13] Since, even in 1930, some lung cancers in males were undoubtedly due to smoking, and since there is no reason to anticipate a sex-linked difference in lung cancer, the female rate probably reflected the true nonsmoker lung cancer rate. Was there a marked underdiagnosis of lung cancer among women in 1930? This is not likely, since the rate increased only slowly until 1960, when the effects of smoking among women since World War II resulted in a continually steeper rise in their lung cancer death rates. Thus, the lung cancer death rates among nonsmokers should be considerably less than the numbers estimated by the various agencies, since one cannot attribute all lung cancer in nonsmokers to radon and its daughters. Evidence that the major fraction of lung cancer in nonsmokers is not associated with radon exposure comes from histologic studies. The predominant histologic diagnosis of lung cancer in nonsmokers is adenocarcinoma (65%), with large-cell carcinoma accounting for about one-half the rest.[14] Adenocarcinoma of the lung is rare in uranium miners; it is the small-cell undifferentiated cell types that predominate and appear to be related to cumulative radiation exposure.[15] Although there may have been some increase recently in indoor radon levels due to the emphasis on energy-efficient homes during the past decade, that is hardly likely to have affected lung cancer death rates in the 1980s.

Taken together, this evidence suggests that the BEIR III estimates are too high to be used in estimates for radon-induced lung cancer in the home setting, perhaps because of the presence in uranium mines of carcinogens other than radon and its daughters or because even high LET radiation, if delivered at low dose rates, as occurs in the homes with elevated radon levels, is less carcinogenic than similar radiation delivered at higher dose rates to miners.

LIMITATIONS OF CASE-CONTROL METHODOLOGY

Can epidemiologic studies permit testing of the validity of the linear-extrapolation hypothesis in estimating effects at low doses and dose rates? The answer is unequivocally no. As Land has pointed out,[16] testing this hypothesis for radiation-induced breast cancer would require a sample size of 100 million women to be certain of an increased cancer incidence following an acute exposure of 1 rem to both breasts at age 35. Such a sample is hardly practical. Therefore, a case-control approach, in which the sample consists of a fixed number of cancer cases and a fixed number of matched noncases or controls, is used. Land has calculated that if this cohort approach were used, only 1,000,000 women would be required to be certain of a radiation effect from 1 rem.[16] Of course, in the case-control approach to evaluation of radiation and other potential carcinogens, a sufficient number of subjects are never included and there is not random selection of cases and controls. Hence, the data presented simply do not have statistical significance, and subtle sources of bias could well account for purported observed effects. For instance, Mac-Mahon [17] has reported that children born after their mothers had received one to six pelvic radiographs (average dose per radiograph was 1 rad) were 42% more likely to die of cancer in the first 10 years of life than were children not irradiated in utero. Using the same case-control method of analysis, MacMahon et al.[18] also reported that drinking 1 to 2 cups of coffee a day introduced a relative risk of 2.6 in developing cancer of the pancreas and further suggested that coffee drinking at this level can account for more than 50% of the cases of pancreatic cancer. However, since coffee drinking is familiar and radiation is not, most people would discount that his case-control analysis proves that such modest coffee drinking is a risk factor for pancreatic cancer, particularly since the effect did not appear to be dose-related in men — the risk factor was the same, 2.6, whether consumption was one to two cups or greater than five cups a day.

There are other reasons for reluctance in accepting his analysis. For instance, the risk factor was twice as great for ex-smokers compared to current smokers drinking one to two cups of coffee per day — a rather unlikely finding, since it is commonly accepted that smoking is a carcinogen or promoter of other carcinogens. The MacMahon et al. report on the association between coffee drinking and cancer of the pancreas[18] is, however, in a sense less flawed than his earlier report on the association between prenatal radiation and early cancer death.[17] In the latter study, there was clearly a bias, in that no account was taken of the fact that the exposed mothers had medical conditions that prompted the diagnostic X-rays.

RADIATION EXPOSURES GREATLY IN EXCESS OF NATURAL BACKGROUND WITH NO OBSERVABLE DELETERIOUS EFFECTS

There are several studies in which increases in malignancies were not observed in spite of radiation exposures considerably greater than those associated with the usual variations in natural background radiation. One large group of subjects with total body exposures in the 5 to 15 rem range consists of patients treated with radioactive iodine, I-131, for hyperthyroidism. As of 1968 it was estimated that 200,000 people in the United States had been so treated, and the number has probably doubled since. A study of 36,000 hyperthyroid patients from 26 medical centers, of whom 22,000 were treated with a single dose of I-131 and most of the rest with surgery, revealed no difference in the incidence of leukemia between the two groups.[19] The average bone-marrow dose was estimated to be about 10 rems, more than half of which was delivered within one week. The follow-up for the I-131-treated group averaged seven years, quite long enough to have reached the peak incidence for leukemia, as had been determined from the Hiroshima-Nagasaki experience.[4] Another follow-up of the hyperthyroid patients three years later continued to reveal no difference between the two groups in their leukemia rates.[20] This study emphasizes the importance of having an appropriate control group.[19] Earlier studies had suggested that the occurrence of leukemia in hyperthyroid patients following I-131 therapy was 50% greater than that of the natural population.[21,22] However, it appears from this study that there is an increased incidence of leukemia in hyperthyroidism, irrespective of the type of treatment.[19]

The feasibility of a large epidemiologic study of the several hundred thousand patients who have been treated with I-131 for hyperthyroidism should be considered. Such a study could answer the question whether general body exposure in the 10 rem range, approximately half of which is delivered in less than one week, is carcinogenic. Since it appears that hyperthyroidism per se may be associated with leukemia, the appropriate control group should be, as in the study of Saenger et al.,[19] patients treated with surgery. However, it may not be possible to obtain an age-matched surgically treated group, since I-131 certainly has become the treatment of choice for definitive therapy. In evaluating whether hyperthyroid patients treated with antithyroid drugs until remission would be suitable as a control group, the potential of these drugs for inducing leukemia must also be considered, since, in some patients, antithyroid drugs are known to produce bone marrow depression.

The question of malignancies induced by diagnostic X-radiation remains a matter of public concern. In addition to the study of MacMahon[17] described earlier, Stewart et al.[23] reported a positive association between acute and chronic myelocytic leukemia and diagnostic X-radiation. That study was based on questionnaires given by physicians to leukemia patients and control subjects. The numbers of diagnostic procedures were then counted and a uniform quantity of radiation dose was assumed for each procedure. All information was based on subjects' recall, not on objective confirmation by medical records. However, a case-control study by Linos et al.[24] of 138 cases of leukemia, which represented all known cases in Olmsted County, Minnesota between 1955 and 1974 and matched controls, revealed that there was no statistically significant increase in the risk of developing leukemia after radiation doses up to 300 rads to the bone marrow when these doses were administered in small doses over long periods of time, as is the case in routine medical care. The difference between the study by Stewart et al.[23] and that of Linos et al.[24] was that in the latter study the exposure to radiation administered for medical reasons had been prerecorded and was documented through careful review of records which had been entered when the radiation exposures had actually occurred, not after the diagnosis of leukemia was made. The Olmsted County experience is quite unique in that virtually all medical care is provided by the Mayo Clinic and one other private medical group practice and that the record keeping and estimations of bone marrow dose are very reliable.

There have been other negative studies in which induction of leukemia as a consequence of exposure to radiation therapy was sought for and not found. Perhaps early studies[25,26] in which no increase in leukemia in women treated for cervical cancer with either intracavitary radium, external radiation, or both were neglected because of incomplete patient follow-up. However, a report of an international collaborative study of more than 31,000 women with cervical cancer, of whom 90% received radiation therapy and 10% did not, revealed that 15.5 cases of leukemia were expected in the irradiated group but only 13 were observed (relative risk = 0.8, 95% confidence limits 0.41.4).[27] In the nonirradiated group, two cases of leukemia were observed, as compared with the 1.0 expected. The follow-up was long enough to have included the period of leukemia peaking observed with the Japanese atom bomb survivors.[4] The consistency of these studies[25-27] would suggest that there is no detectable leukemogenic effect in patients with cervical cancer following radiotherapy. The cohort size of this study is quite comparable to those of the Court-Brown and Doll studies, showing increased incidence of leukemia in patients irradiated for ankylosing spondylitis.[28,29] It does remain a mystery as to why radiotherapy would appear to be leukemogenic in one disease and not in another when the therapeutic doses are in the same range, although not delivered to identical regions of the body.

It is commonly accepted that early radiation workers had an increased incidence of malignancies. For the most part, their radiation exposures cannot be estimated. The classical picture of the Curies working in their shed for four years while separating and purifying radium and polonium is one that will never be repeated. It is not surprising that Marie Curie died from aplastic anemia, probably secondary to the radiation exposure she received in her laboratory and during her experiences in World War I, when she personally provided X-ray services just behind the front lines, trained X-ray technicians, and installed 200 radiologic rooms. What is surprising is that she did not die until 1934 at the age of 66, in spite of cumulative exposures that must have been thousands of rems.

What is the radiation-related incidence of malignancies in more recent radiation workers with lesser exposures? A report of the mortality from cancer and other causes among 1338 British radiologists who joined radiologic societies between 1897 and 1954 revealed that in those who entered the profession before 1921, the cancer death rate was 75% higher than that of other physicians, but that those entering radiology after 1921 had cancer death rates comparable to those of other professionals.[30] Although the exposure of the radiologists was not monitored, it is estimated that those who entered between 1920 and 1945 could have received an accumulated whole-body dose of the order of 100 to 500 rem.[30]

Another large group of radiation workers was also studied. These were men in the U.S. armed services who were trained as radiology technicians during World War II and who subsequently served in that capacity for a median period of 24 months. Description of their training included the statement that "During the remaining two hours of this period the students occupy themselves by taking radiographs of each other in the positions taught them that day."[31] In this report it was noted that the students did not receive a skin erythema dose nor did they show a drop in white count. These monitoring procedures are insensitive to acute doses less than 100 rem. Although the cumulative exposures of these radiology technicians were not monitored, the radiation exposure of technologists at a more modern installation, Cleveland Clinic, was monitored in 1953 and found to be in range of 5 to 15 rem/yr.[32] Army technologists a decade earlier probably received as much as 50 rem or more during their training and several years of service. Yet, a 29-year follow-up of these 6500 radiology technicians revealed no increase in malignancies when compared with a control group of similar size consisting of Army medical, laboratory, or pharmacy technicians.[33,34]

POTENTIAL FOR INCREASE IN MALIGNANCIES ASSOCIATED WITH TESTING OF NUCLEAR WEAPONS

As discussed earlier, by 1978 the 80,000 Japanese atomic bomb survivors, who were estimated to have received about 25 rem, manifested less than a 6% increase in the natural incidence of malignancies that could be attributable to radiation.[4] Even were the linear extrapolation valid and if there were no dose rate effect, this would mean that for those receiving 2.5 rem of total body radiation exposure, the increase in malignancies which would be radiation-related would be less than 1% — an effect that would not be measurable above the natural variation in incidence of malignancies. Therefore, whether or not to expect an increase in malignancies associated with testing of nuclear weapons would depend on the validity of estimates of radiation exposure. There are some who hold that the atomic bomb survivor studies underestimate radiation-related malignancies because the trauma of the bomb blast destroyed the weakest people and the survivors should be more resistant to malignant disease.[35] Perhaps it might be of interest to determine whether the survivors of the Tokyo or Dresden fire bombings had a lower malignancy rate than others living in unbombed cities, but until now no one has taken the Morgan thesis[35] sufficiently seriously to have suggested such epidemiologic studies.

Considerable publicity has been given to problems of the so-called Atomic Veterans. Caldwell et al.[36] reported an increased incidence of leukemia among 3200 men who participated in Operation Smoky, a nuclear explosion at the Nevada Test Site in 1957. Stimulated by this report, the Medical Follow-up Agency of the National Research Council studied the mortality and causes of death of a cohort of 46,186 participants, about 1/5 of the total number of participants in one or more of five atmospheric nuclear tests.[37] The reanalysis confirmed that among the 3500 participants at Operation Smoky the standardized mortality ratio (SMR) for leukemia was 2.5; that is, there were 10 observed leukemia deaths and only 3.97 expected. Only one of those 10 received an exposure in excess of 3 rem. For all other cancers the SMRs were less than 1.0. It is of particular interest that among participants in Operation Greenhouse at a Pacific test site in 1951, with a cohort size of almost 3000, the expected leukemia mortality was 4.43, yet only one was observed, for a SMR of 0.23. For the other malignancies, where the numbers involved are much larger, the SMRs are in the range of 0.7 to 0.9. If we examine the entire cohort of 46,000, we find the SMR for all malignancies is 0.84 and for leukemia it is 0.99. The excess SMR for leukemia at Operation Smoky and the equivalently decreased SMR at Operation Greenhouse are typical aberrations attributable to small-number statistics. Could one have expected an increase in leukemia at Operation Smoky? Since only one of the veterans with leukemia was reported to have received more than 1 rem, the probability of observing a true increase in leukemia would require a gross underestimate of the radiation dose received by the participants. However, a committee, chaired by Dr. Merril Eisenbud for the National Research Council, reviewed the methods used to assign radiation doses to service personnel at nuclear tests and concluded that the methods were reasonably sound but that doses assigned to the test participants were probably somewhat higher, not lower, than the actual doses received.[38] This report also reviewed a number of studies that estimated radiation exposure from internally deposited radionuclides and concluded that these did not add significantly to the external exposure.

There have also been reports of increases in malignancies among civilians exposed to fallout from nuclear testing. In 1979, Lyon et al.[39] reported that leukemia mortality in children was increased in those counties in Utah receiving high levels of fallout from the atmospheric nuclear testing conducted in 1951–1958 compared to the mortality in low-fallout counties and in the rest of the United States. Let us examine the original data. In Figure 1 are shown the mortality rates for leukemia and for all cancers, including leukemia, for children in high- and low-fallout counties in Utah. The 1944–1950 period represents the prefallout control low-exposure cohort. The 1951–1958 group was considered to be the high-exposure cohort, that is, those born during the period of maximum above-ground nuclear bomb testing in Nevada. The second low-exposure cohort was the group born after most, but not all, of the above-ground testing had ended. From a perusal of Lyon's data, it could reasonably be concluded that on the average, during the entire 30-year period, the high-fallout counties might have had a lower incidence of leukemia than the low-fallout counties, but that the uncertainties in the determinations are so large that one cannot reliably conclude whether or not there is a trend. If one considers the sum of childhood malignancies (leukemia plus other cancer deaths), there appears to be a generally downward trend, with the drop in the high-fallout counties being somewhat greater than in the low-fallout counties, although if the standard deviations had been included the differences would not have been significant. The news headlines following interviews with Dr. Lyon would have been less sensational if he had stated that his data had shown no relation between the totality of childhood malignancies and the atomic tests of the 1950s rather than selectively reporting an inconclusive study of the relation between leukemia and fallout.

Lyon's paper in the *New England Journal of Medicine*[39] was criticized in the same and later issues of the journal by several biostatisticians.[40–42]

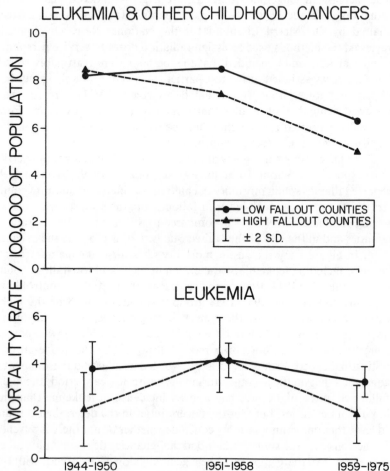

Figure 1. Total childhood malignancies including leukemia (top) and childhood leukemia (bottom) in high-fallout and low-fallout regions of Utah. Data reproduced in graphical form from J. L. Lyon, M. R. Klauber, J. W. Gardner, and K. S. Udall.[39]

In general, their criticisms were related to the apparent underreporting or misdiagnosis in the earlier cohort and to errors in small-sample analysis. For instance, Bader presented a year-by-year listing of leukemia cases in Seattle–King County, which has a larger population than the southern Utah counties, and noted that there were only two cases in 1959 and 20 in 1963 among the 217 cases reported from 1950 to 1972.[41] Thus, a tenfold difference in annual incidence rates when the number of cases is small simply represents statistical variation. Although the yearly distribution of leukemia cases has never been reported in any of Lyon's papers or in the associated publicity, it was tabulated in the General Accounting Office (GAO) report "Problems in Assessing the Cancer Risks of Low-Level Ionizing Radiation Exposure."[43] It is of interest that the so-called excess of leukemia cases reported by Lyon et al.[39] was due to a clustering of 13 cases in 1959 and 1960 (Table 2). In fact, 22 of the 32 leukemia cases occurred in 1951–1960, i.e., during the first 10 years of testing. Since there is a several-year latency period between radiation exposure and induction of leukemia, were the excess leukemia deaths a consequence of nuclear testing in the 1950s, they are more likely to have occurred after 1960 rather than before.

Furthermore, a new estimation of external radiation exposure of the Utah population based on residual levels of Cs-137 in the soils has shown that the mean individual exposure in what Lyon deemed to be the "high-fallout counties" was 0.86 ± 0.14 rad compared to 1.3 ± 0.3 rad in the "low-fallout counties."[44] Even in Washington County, the region in which the fallout arrived the earliest (less than five hours after the test), the estimated exposure to its population of 10,000 averaged only 3.5 ± 0.7 rads—quite comparable to natural background radiation in that region over a 20-year period. Thus on the basis of the Japanese experience the exposure from fallout was too low for an increase in leukemia to be expected, and a careful perusal of Lyon's data would suggest that none was found.

Measurement of residual radioactivity in the soil permits reconstruction of the external gamma-ray exposure of the population downwind from the nuclear test sites. However, there has also been concern with inhaled or ingested radionuclides. Therefore, a 10-year program was initiated in 1970 to determine levels of radionuclides in adults and children from families residing in communities and ranches surrounding the Nevada Test Site.[45] Monitoring was done by whole body counting, which would determine Co-60, I-131, and Cs-137, and by testing of urine for Pu-239 and Pu-23. I-131 was included in the monitoring because the underground Baneberry test of 1970 was vented and there was some concern about I-131 contamination of the milk supply. Monitoring revealed that the estimated thyroidal dose to members of a family using

Table 2.   Year of Bomb Testing in Nevada[a] and Year of Death From Leukemia for the Southern Utah High-Exposure Cohort[b]

	Test Yield (kilotons)	Number of Deaths Southern Utah High-Exposure Cohort
51	112	0
52	104	3
53	252	2
54	—	0
55	167	2
56	?	2
57	343	0
58	38	0
59	—	7
60	—	6
61	—	2
62	101	1
63	<20	1
64	<20	0
65	<80	1
66	<60	1
67	20–200	2
68	<60	1
69	20–200	1

[a]Source: U.S. Comptroller General.[43]
[b]From data supplied by Dr. Lyon for the Comptroller General's report.[43]

the contaminated milk was about 6 mrem, about 1/10,000 of the thyroidal dose routinely received by the several million people who had received diagnostic doses of I-131 for testing of thyroid function between 1948 and 1968. Cs-137 was the major fission product found in the offsite population and its concentration was similar to that found in people living elsewhere in the United States. No clear difference was found between persons living in the windward and leeward sides of the Nevada Test Site.[45]

In view of the failure to demonstrate significant external or internal radiation doses resulting from fallout from nuclear testing, what then could account for the remarkably high incidence of many types of cancer downwind from the Nevada test site reported by Johnson?[46] Johnson reports from a series of personal or oral communications that doses received by residents were in the range of hundreds to thousands of rads. He does not reference or discuss the earlier published scientific reports demonstrating the failure to detect significant contamination of the people or soil in the fallout areas.[44,45] However, he does reference a "News and Comment" article in *Science* [47] that reports on a lawsuit against the government of the United States in which allegations were made that the government was grossly underestimating human radiation exposure following nuclear bomb testing. Even were there some underestimation of such exposure, it is most remarkable that through 1975 there were 288 cancer deaths in the so-called exposed group (4125 people) Johnson reports compared to 179 expected,[46] whereas the Japanese experience was only an extra 250 cancer deaths over the 4500 expected by 1978 in a population of 82,000.[4] The Johnson report[46] is certainly inexplicable.

USEFULNESS OF ANIMAL STUDIES IN PREDICTION OF HUMAN RADIATION-INDUCED CARCINOGENESIS

Animal studies have certain advantages: the animals are inbred and are not subject to the genetic and environmental variability of a human population; at present it is possible to expose animals but not humans to graded radiation doses at different dose rates. The inherent limitation of such studies is that it would be enormously expensive to maintain the large groups of animals that would be required to evaluate effects at truly low doses and dose rates. The conclusion of many studies of different tumors in different animals is that for a given total dose there was generally decreased tumorigenesis when the radiation was delivered at a lower dose rate, but that the reduction factor was dependent on both the tumor type and the species of animal.[48] None of these studies has been performed at truly low dose rates. For instance, the studies by Ullrich and Storer[49] on tumorigenesis in mice revealed that when Cs-137 gamma-ray irradiation is delivered at 8.3 rad/day, i.e., 25,000 times natural background, there is a threshold of about 50 rads before an increased incidence of ovarian tumors or thymic lymphomas is observed. The threshold appeared to be no more than a few rads when the irradiation dose rate was 45 rad/min. Studies at very low dose rates—for instance, at about 100 times natural background—would require an enormous number of animals and are really not practical.

GENETIC AND MUTAGENIC EFFECTS OF RADIATION

In addition to concerns with carcinogenesis, there is considerable fear of the risks of genetic effects from radiation. This was considered extensively in the BEIR III report[11] and it was concluded that since radiation-induced transmitted genetic effects have not been demonstrated in man, estimates of genetic risks must be based on laboratory data obtained at high dose rates. Schull et al.[50] have concluded on the basis of studies of the children born to survivors of the Hiroshima-Nagasaki bombings, that the estimated doubling dose for genetic changes would be about 150 rem, a value some fourfold higher than the results from experimental studies on mice. Furthermore, this represents simply an estimate, since they reported that in *no* instance was there a statistically significant effect of parental exposure. It should be noted also that many investigators have found that for the same total dose, chronic irradiation in mice is about threefold less effective than acute irradiation. This would effectively raise the doubling dose for genetic changes in man to about 500 rem of prolonged low dose rate exposure. Furthermore, since none of the studies in mice was performed at truly low dose rates, one cannot really determine what the doubling rate would be for background radiation.

Recently Gevertz et al.[51] have attempted to evaluate the contribution to the spontaneous mutation rate in *E. coli* due to the K-40 which contributes about one-third to their natural background radiation. These investigators compared the spontaneous mutation rate of *E. coli* in chemostat cultures of normal isotopic composition with that of cultures identical in all respects except for the use of potassium, from which K-40 had been removed or had been enhanced by fiftyfold or thousandfold enrichment. On the basis of more than 40 such chemostat experiments, each typically requiring from 10 days to two weeks, the failure to observe differences in mutation rates in spite of large differences in radiation exposure led these investigators to conclude that it was unlikely that natural K-40 had been an important factor as a primordial gene irradiator in evolution. Previously, Moore and Sastry had conjectured on theoretical grounds that "intracellular K-40 had played a significant role as a mutagenic agent in evolution."[52] Although a single study[51] does not close the issue, the Moore-Sastry hypothesis remains to be validated.

HYPOTHESIS VERSUS OBSERVABLE FACTS

The GAO report concluded that "there is as yet no way to determine precisely the cancer risks of low-level ionizing radiation exposure, and it is unlikely that this question will be resolved soon."[1] The question as to whether there exists a threshold below which radiation effects in man do not occur should again be addressed. One can develop a tenable model that would be consistent with such a threshold. Since human beings are more than 75% water, low-LET ionizing radiation is largely absorbed in the water with production of free radicals. Thus, many of the biochemical changes initiated in the cell and, in particular, damage to cellular

DNA are probably a consequence of the indirect action of the products of water radiolysis. If molecules which scavenge radicals and which are normally present in tissue exceed the concentration of free radicals generated at low dose rates, there may well be no initiating event, i.e., damage to DNA. The threshold could be the dose rate at which the free radicals overwhelm the scavengers. The threshold is likely to be dependent on the animal species and the specific tissue of concern. Such a hypothesis is consistent with the marked dose rate effect observed in animal studies.

CONCLUSIONS

From the studies described in this review, it is evident that epidemiologic studies cannot produce meaningful data about the existence of a threshold for radiation effects. Molecular and cellular studies may or may not give some insight about molecular or cellular effects, but cannot answer important questions about repair mechanisms in the intact animal or man when radiation is delivered at low dose rates. At present there are no really good ideas that would permit a breakthrough in the field of evaluation of potential effects of low-level radiation delivered at a low dose rate.

The disagreement in the low-level radiation field is about hypothesis — not observable facts. One could not determine the validity of Newton's laws at subatomic dimensions until the tools became available. However, in that case there was no need to make policy decisions based on extrapolation. In the case of low-level radiation effects, public policy decisions need to be made in the absence of scientific evidence. It should be appreciated that these are arbitrary decisions based on philosophy, not fact, and may well change because of political or other considerations.

In conclusion, a quotation from the National Council on Radiation Protection (NCRP) Report 23 on Radiation Protection Philosophy is relevant.

The indications of a significant dose rate influence on radiation effects would make completely inappropriate the current practice of summing of doses at all levels of dose and dose rate in the form of total person-rem for purposes of calculating risks to the population on the basis of extrapolation of risk estimates derived from data at high doses and dose rates. . . . The NCRP wishes to caution governmental policy-making agencies of the unreasonableness of interpreting or assuming 'upper limit' estimates of carcinogenic risks at low radiation levels, derived by linear extrapolation from data obtained at high doses and dose rates, as actual risks, and of basing unduly restrictive policies on such an interpretation or assumption.

Undue concern, as well as carelessness with regard to radiation hazards, is considered detrimental to the public interest.

REFERENCES

1. Comptroller General of the United States, Report to the Congress of the United States, "Problems in Assessing the Cancer Risks of Low-Level Ionizing Radiation Exposure" (EMD-81-1), Volume 1 (1981).
2. "Natural Background Radiation in the United States" (Washington, D.C.: National Council on Radiation Protection [NCRP Report No. 45], 1975).
3. High Background Radiation Research Group, China, "Health Survey in High Background Radiation Areas in China," *Science* 209:877–880 (1980).
4. Kato, H., and Schull, J., "Cancer Mortality among Atomic Bomb Survivors, 1950–78," *Radiation Res.* 90:395–432 (1982).
5. Frigerio, N. A., and Stowe, R. S., "Carcinogenic and Genetic Hazard from Background Radiation," in *Biological and Environmental Effects of Low-Level Radiation* (Vienna: International Atomic Energy Agency, 1976), pp. 385–393.
6. Barcinski, M. A., Abreu, M.-D. A., DeAlmeida, J. C. C., Naya, J. M., Fonseca, L. G., and Castro, L. E., "Cytogenetic Investigation in a Brazilian Population Living in an Area of High Natural Radioactivity," *Am. J. Hum. Genet.* 27:802–806 (1975).
7. Cullen, T. L., "Dosimetric and Cytogenetic Studies in Brazilian Areas of High Natural Activity," *Health Physics* 19:165–166 (1970).
8. Gruneberg, H., Bains, G. S., Berry, R. J., Riles, L., Smith, C. A. B., and Weiss, R. A., "A Search for Genetic Effects of High Natural Radioactivity in South India" (London: H.M.S.O. [Medical Research Council Special Report Series No. 307], 1966).
9. Gopal-Ayengar, A. R., Sundaram, K., Mistry, K. B., Sunta, C. M., Nambi, K. S. V., Kathuria, S. P., Basu, A. S., and David, M., "Evaluation of the Long-Term Effects of High Background Radiation on Selected Population Groups on the Kerala Coast," in *Peaceful Uses of Atomic Energy* (1972), pp. II:31–51.
10. Ahuja, Y. R., Sharma, A., Nampoothiri, K. U. K., Ahuja, M. R., and Dempster, E. R., "Evaluation of Effects of High Natural Background Radiation on Some Genetic Traits in the Inhabitants of Monazite Belt in Kerala, India," *Hum. Biol.* 45:167–179 (1973).
11. BEIR III Report, "The Effects on Populations of Exposure to Low Levels of Ionizing Radiation" (Washington, DC: Committee on the Biological Effects of Ionizing Radiations, Assembly of Life Sciences, National Academy of Sciences, 1980).
12. *Ca:A Cancer Journal for Clinicians* (American Cancer Society, 1986), 36:1.
13. *Ca:A Cancer Journal for Clinicians* (American Cancer Society, 1983), 33:1.

14. Garfinkel, L., Auerbach, O., and Jouber, L., "Involuntary Smoking and Lung Cancer: A Case-Control Study," *J. Nat. Cancer Inst.* 75:463–469 (1985).

15. Soccomanno, G., Archer, V. E., Auerbach, O., Kuschner, M., Saunders, R. P., and Klein, M. G., "Histologic Types of Lung Cancer among Uranium Miners," *Cancer* 71:158–523 (1971).

16. Land, C. E., "Estimating Cancer Risks from Low Doses of Ionizing Radiation," *Science* 209:1197–1203 (1980).

17. MacMahon, B., "Prenatal X-Ray Exposure and Childhood Cancer," *J. Nat. Cancer Inst.* 28:1133–1191 (1962).

18. MacMahon, B., Yen, S., Trichopoulos, D., Warren, K., and Nardi, G., "Coffee and Cancer of the Pancreas," *New England J. Med.* 304:630–633 (1981).

19. Saenger, E. L., Thoma, G. E., and Tompkins, E. A., "Incidence of Leukemia Following Treatment of Hyperthyroidism," *JAMA* 205:855–862 (1968).

20. Saenger, E. L., Tompkins, E., and Thoma, G. E., "Radiation and Leukemia Rates," *Science* 171:1096–1098 (1971).

21. Pochin, E. E., "Leukemia Following Radioiodine Treatment of Thyrotoxicosis," *Brit. Med. J.* 2:1545 (1960).

22. Werner, S. C., Gittleshon, A. M., and A. B. Brill, "Leukemia Following Radioiodine Therapy of Hyperthyroidism," *JAMA* 177:646–648 (1961).

23. Stewart, A., Pennybacker, W., and Barber, R., "Adult Leukemias and Diagnostic X-Rays," *Brit. Med. J.* 2:882–890 (1962).

24. Linos, A., Gray, J. E., Orvis, A. L., Kyle, R. A., O'Fallon, W. M., and Kurland, L. T., "Low Dose Radiation and Leukemia," *New England J. Med.* 302:1101–1105 (1980).

25. Simon, N., Bruger, M., and Hayes, R., "Radiation and Leukemia in Carcinoma of the Cervix," *Radiology* 74:905–911 (1975).

26. Hutchinson, G. B., "Leukemia in Patients with Cancer of the Cervix Uteri Treated with Radiation" [a report covering the first five years of an international study], *J. Nat. Cancer Inst.* 40:951–982 (1968).

27. Boice, J. D., and Hutchinson, G. B., "Leukemia in Women Following Radiotherapy for Cervical Cancer" [10-year follow-up of an international study], *J. Nat. Cancer Inst.* 65:115–129 (1980).

28. Court-Brown, W. M., and Doll, R., "Leukemia and Aplastic Anemia in Patients Irradiated for Ankylosing Spondylitis" (London: Medical Research Council Special Reports, 1957), 295:1–135.

29. Court-Brown, W. M., "Mortality from Cancer and Other Causes after Radiotherapy for Ankylosing Spondylitis," *Brit. Med. J.* 2:1327–1332 (1965).

30. Smith, P. G., and Doll, R., "Mortality from Cancer and All Causes among British Radiologists," *Brit. J. Radiol.* 54:187–194 (1981).

31. McCaw, W. W., "Training of X-Ray Technicians at the School for Medical Department Enlisted Technicians," *Radiol.* 42:384–388 (1944).

32. Geist, R. M., Jr., Glasser, O., and Hughes, C. R., "Radiation Exposure Survey of Personnel at the Cleveland Clinic Foundation," *Radiology* 60:186–191 (1953).

33. Miller, R. W., and Jablon, S., "A Search for Late Radiation Effects among Men Who Served as X-Ray Technologists in the U.S. Army during World War II," *Radiology* 96:269–274 (1970).

34. Jablon, S., and Miller, R. W., "Army Technologists: 29-Year Follow-Up for Cause of Death," *Radiology* 126:677-679 (1978).

35. Morgan, K. Z., "Appreciation of Risks of Low Level Radiation vs. Nuclear Energy," Comments, *Mol. & Cell. Biophys.* 1 (No.l): 41–55 (1980).

36. Caldwell, G. G., Kelley, D. B., and Heath, C. W., Jr., "Leukemia among Participants in Military Maneuvers at a Nuclear Bomb Test," *JAMA* 244:1575–1578 (1980).

37. Robinette, C. D., Jablon, S., and Preston, T. L., "Studies of Participants in Nuclear Tests," report to the National Research Council, in *Mortality of Nuclear Weapons Test Participants* (Washington, DC: National Academy Press, 1985).

38. Board on Radiation Effects Research, Commission on Life Sciences, "Review of the Methods Used to Assign Radiation Doses to Service Personnel at Nuclear Weapons Tests" (Washington, DC: National Academy Press, 1985).

39. Lyon, J. L., Klauber, M. R., Gardner, J. W., and Udall, K. S., "Childhood Leukemias Associated with Fallout from Nuclear Testing," *New England J. Med.* 300:397–402 (1979).

40. Land, C. E., "The Hazards of Fallout or of Epidemiologic Research?", *New England J. Med.* 300:431–432 (1979).

41. M. Bader, "Leukemia from Atomic Fallout," *New England J. Med.* 300:1491 (1979).

42. Enstrom, J. E., editorial, *New England J. Med.* 300:1491 (1979).

43. Comptroller General of the United States, "Problems in Assessing the Cancer Risks of Low-Level Ionizing Radiation Exposure: Report to the United States Congress" (EMD-81-1) (Washington, DC: U.S. General Accounting Office, 1981), Volume 2, pp. xviii-43.

44. Beck, H. L., and Krey, P. W., "External Radiation Exposure of the Population of Utah from Nevada Weapons Tests" (DOE/EML401 [DE82010421]), National Technical Information Service, U.S. Department of Energy, NY (1982), p. 19.

45. Patzer, R. G., and Kaye, M. E., "Results of a Surveillance Program for Persons Living around the Nevada Test Site, 1971 to 1980," *Health Physics* 43:791–801 (1982).

46. Johnson, C. J., "Cancer Incidence in an Area of Radioactive Fallout Downwind from the Nevada Test Site," *JAMA* 251:230–236 (1984).

47. Smith, R. J., "Atom Bomb Tests Leave Infamous Legacy," *Science* 218:266–269 (1982).

48. "Tumorigenesis in Experimental Laboratory Animals" (Washington, DC: National Council on Radiation Protection and Measurements [Report 64], 1980).

49. Ullrich, R. L., and Storer, J. B., "Influence of Dose, Dose Rate and Radiation Quality on Radiation Carcinogenesis and Life Shortening in RFM and BALB/c Mice," in *Late Biological Effects of Ionizing Radiation* (Vienna: International Atomic Energy Agency [IAEA/STI/PUB/489], 1978), Volume 2, p. 95.

50. Schull, W. J., Otake, M., and Neel, J. V., "Genetic Effects of the Atomic Bombs: A Reappraisal," Science 213:1220–1227 (1981).

51. Gevertz, D., Friedman, A. M., Katz, J. J., and Kubitschek, H. E., "Biological Effects of Background Radiation: Mutagenicity of K-40," *Proc. Nat. Acad. Sci. U.S.* 82:8602–8605 (1985).

52. Moore, F. D., and Sastry, K. S. R., "Intracellular Potassium: K-40 as a Primordial Gene Irradiator," *Proc. Nat. Acad. Sci. U.S.* 79:3556-3559 (1982).

CHAPTER 13

Safer Than Sleeping with Your Spouse— The West Valley Experience

John M. Matuszek

INTRODUCTION

The Low-Level Radioactive Waste Policy Act of 1980 (P. L. 96–573) makes each state responsible for the disposal of all low-level radioactive waste generated within the state, except for that from certain federal operations. Most of the low-level radioactive waste in the United States is generated in states with humid climates, making it likely that most new low-level radioactive waste burial sites will find it necessary to cope with water infiltration, as was the case at the now closed commercial burial site at West Valley, New York (see Figure 1).

A variety of interpretations have been provided for the many environmental measurements made on and around the burial site at West Valley. This chapter provides a summary of various studies performed at the site by several federal and state agencies. The discussion is organized in the form of answers to six questions: Why did water enter the burial trenches? How were the radionuclides transported out of the trenches? Where did the radionuclides go? What are the regulatory and health implications? Are the doses from West Valley really trivial? and Can or should more be done to reduce effluents?

WHY DID WATER ENTER THE BURIAL TRENCHES?

Water entered through the trench covers partly as a result of poor site management by the then manager of the waste burial site, Nuclear Fuels Services, Inc. (NFS) and partly from poor waste management both by NFS and by the generators of the radioactive waste shipped there.[1,2] Records

Low-Level Radioactive Waste Regulation: Science, Politics, and Fear, Michael E. Burns, Ed., © 1988 Lewis Publishers, Inc., Chelsea, Michigan – Printed in USA.

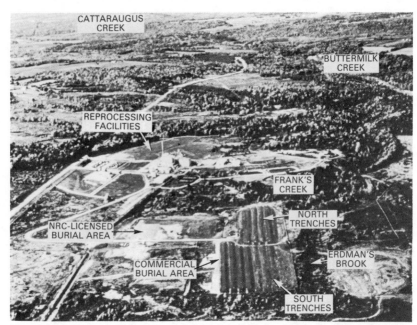

Figure 1. An aerial photograph of a portion of the 3500-acre Western New York Nuclear Services Center (WNYNSC), West Valley, NY, shows the trenches of the commercial burial area (center foreground) as a series of long mounds aligned approximately 30 degrees west of north. A narrow road divides the commercial area into two sections—the so-called "north trenches" and "south trenches." The reprocessing plant and associated structures are approximately centered in the photo, while the NRC-licensed burial area appears as the large, cleared (light) area south of the main reprocessing structure and just west of the commercial area's north trenches. Frank's Creek, which appears as a wooded (dark) area between the burial sites and the main reprocessing facilities, forms a 40-ft deep ravine at the point where its easterly flow carries it past the north trenches. It joins Erdman's Brook (appearing as trees [dark areas] just south and east of the burial mounds) to flow northward into Buttermilk Creek. The northwesterly flow of Buttermilk Creek carries it into Cattaraugus Creek (seen at the top of the photo), which eventually reaches Lake Erie some 35 miles to the west and well out of the range of this photograph (Photo courtesy of N.Y. State Geological Survey).

maintained at the site show that NFS found it necessary every spring to make extensive repairs to the trench covers, to fill large areas which had collapsed into the trenches, to seal sinkholes into which rainwater was running and to regrade and reseed the mounded covers. Many of the collapsed covers or sinkholes occurred from large voids created among the emplaced wastes. Repairs would continue through summer and fall of each year. Staff of the New York State Department of Environmental Conservation (DEC), in their inspections of the site, confirmed the failure of the covers and the DEC files contain an extensive record of that agency's efforts to force NFS to remedy the situation.[1] Studies by the U. S. Geological Survey (USGS) have confirmed that increases in trench-water levels were directly and immediately in response to specific rainfall events, i.e., there was no delay following a rain in the increase of the water level of a trench as there might have been if the water had infiltrated through groundwater flow.[3-5] In several file reports (summarized in reference 6), the New York State Geological Survey (NYSGS), as lead agency for the studies at West Valley, has shown that the site is geohydrologically sound for the purpose intended, but that its technical features were violated by the poor management practices of NFS.[1,2]

Failure to remedy the water infiltration problem cannot be seen as solely the fault of NFS, however. Waste was to have been limited to solid waste, but no requirements were placed on the density of the waste. Since most institutional waste generators lacked the capability to incinerate their combustible waste, many waste drums would arrive at the site 30% to 50% empty. When the degradable contents underwent aerobic and anaerobic biological decomposition, the empty space in the trenches would increase to 70% or more of capacity. Without adequate bulk underneath to furnish support, the integrity of the trench caps could not be maintained. The large volume of the void space, the large quantities of escaping gases, and the direct connection between the trench and the surface of the covers were measured and reported by the NYSGS and State Department of Health (DOH).[6-12]

The conclusions concerning water infiltration are: trenches at West Valley filled, because water entered through fractures, collapsed sections, and sinkholes in the covers; trench covers consisting only of soil cannot be expected to maintain integrity against water infiltration unless sufficient support is maintained beneath them; and compaction of degradable waste will *not* solve the problem of in situ decomposition and creation of large voids.

HOW WERE RADIONUCLIDES TRANSPORTED OUT OF THE TRENCHES?

Early in the studies at West Valley it became apparent that the practice by waste generators of adsorbing radioactive liquid waste on vermiculite (the base for "kitty litter") for burial as "solid waste" contaminated the trenches as the drums were being buried. Even the so-called "clean fill" which was used to cover the buried wastes and to cap the trenches was contaminated.[2] A review of NFS and generator practices revealed that, despite regulations which prohibited the burial of liquid wastes at West Valley, burial of adsorbed liquids was an accepted practice. Thus, precipitation which entered the open trenches during burial operations was immediately as contaminated as the leachate which eventually overflowed the trenches in 1975 (see Figure 2). NFS pumped the contaminated rainwater into the local streams and also onto the surface of the site. The surficial soil used to cover the trenches was therefore contaminated right from the beginning, partly from direct discharge of contaminated rainwater onto the surface of the site and partly from the mixing of contaminated water with the clay being pushed to the surface by the bulldozer as it extended each open trench to bury more waste.[2]

Whether radionuclides would be transported through the groundwater system at West Valley was answered in studies by the USGS.[3-5] Using standard geophysical techniques, the hydraulic conductivity of the clay was measured, along with the direction of flow. All of the subsurface flow three or four feet beyond the walls of the trenches was found by USGS to be downward at a rate of approximately one inch per year.[5] Since the clay layer is 90-ft thick under the burial trenches, USGS estimated that approximately 1000 years would be required before the downward movement of the water would change to lateral movement toward Buttermilk Creek. Yet another 1000 to 4000 years were calculated as the time before a molecule of water from one of the burial trenches would find its way into Buttermilk Creek, the first point of contact with the biosphere. Because of the sorptive capacity of the clay, most radionuclides flow much more slowly than the water and few would ever escape through downward flow. Measurements confirmed that transport of all radionuclides except tritium and carbon-14 (^{14}C) by water flowing through the clay was so severely restricted that it was unmeasurable within the limits of the drilling techniques used, which were in turn state-of-the-art methods developed by USGS for this purpose.[13]

The results of the water-flow study and the radionuclide-transport studies, showed that the only radionuclide which might escape through the groundwater system was ^{14}C, migrating as some organic species at half the rate of water. The concentration of ^{14}C in groundwater is greatly reduced by retention of ^{14}C-containing compounds on the trench walls and floor, by exchange with aged carbonate in the soil through which it passes, by biodegradation to gaseous $^{14}CH_4$ and $^{14}CO_2$ which vent through the trench covers, and by radioactive decay during the 4,000- to 10,000-year journey to the surface streams. The concentration of waste-related ^{14}C in groundwater reaching Buttermilk Creek would be so low that it could not be distinguished from ambient ^{14}C which would be

Figure 2.   Burial of radioactive waste at West Valley. An example is provided of how the site operator permitted precipitation in an active trench to come in contact with adsorbed liquid waste in drums (foreground). "Trash-type" waste, which is barely visible behind the stacked drums, was dumped very loosely into the trenches, providing immediate contact with collected precipitation, as well as a poor support for covering material redistributed from the mound seen at the upper right corner of the photograph. A section of pipe used to remove contaminated precipitation from the open trench can be seen lying on the mound of fill material. Although drums are shown neatly stacked in this photo, such care was not always practiced. (Photo courtesy of N.Y. State Geological Survey).

present in surficial soils from natural production of ^{14}C in the atmosphere.

A U. S. Department of Energy (DOE) overflight during 1979 showed a "prong" of cesium-137 (^{137}Cs) on the ground to the northwest of the reprocessing plant,[14] raising concern that an underground seep from the commercial burial site might be the source. However, NFS in 1968 reported to the U. S. Atomic Energy Commission an atmospheric release from the reprocessing plant which contaminated the ground to the northwest of the plant. Measurements reported in DEC's 1971 annual report confirmed that the ^{137}Cs-prong already existed in 1971.[15] Data from an unpublished 1983 study by Westinghouse showed that the ^{137}Cs in the prong was only located on the surface; i.e., the ^{137}Cs had not been transported upward from some depth. Studies by DOH and USGS show that ^{137}Cs has not migrated past the interior surfaces of the bottoms of the trenches[7,13], a finding which is consistent with a body of literature which shows that ^{137}Cs becomes fixed on any clay particles with which it comes in contact.[16] Furthermore, a 40-ft deep ravine between the burial site and the ^{137}Cs prong extends down to bedrock; and the tops of the burial trenches are at an elevation of 1380 to 1390 feet, while the "prong" is at an elevation of 1450 to 1550 feet. Since water cannot run uphill unless an artificial head is created (i.e., by pumping), it is difficult to envision how water from the trenches could transport ^{137}Cs downward beneath a 40-ft deep ravine, through bedrock over which a stream flows, thence to a ridge a half-mile away and uphill some 60 to 170 feet. Scholle derived the same conclusion independently.[17]

Therefore, overflow of the trenches by water infiltrating through the covers to carry radioactive leachate to surface streams and biodegradation of tritium- and ^{14}C-containing wastes to vent radioactive gases through the covers are the primary mechanisms for radionuclides to escape the trenches.

WHERE DID THE RADIONUCLIDES GO?

From 1967 to 1982, DOH and DEC have operated a water-sampling station on Cattaraugus Creek just above Springville Dam, at a point where all surface drainage from the commercial burial site passes as it flows downstream. The sampler is a "continuous collector," sampling every few minutes a portion of the water flowing by it; the container is removed weekly and shipped to the DOH for analysis. Because the dam serves as a large holding and mixing basin, the radionuclides which enter from the site down Buttermilk Creek into Cattaraugus Creek about two miles above the dam are well mixed with the creek water by the time they are collected in the continuous sampler. The results of measurements made on the weekly-composite samples have been reported each year in DEC's Annual Report of Environmental Radiation as "maximum," "minimum," and "average" values for each 52-week period. The annual average values for tritium and strontium-90 (^{90}Sr) are shown in Figure 3. After 1981, the results are for "grab-samples."

When the reprocessing plant at the site ceased operation in November 1971, the concentrations of tritium and ^{90}Sr decreased markedly during 1971 and 1972. The ^{90}Sr concentrations began to decrease a year earlier than did those for tritium, because NFS began during May 1971 to

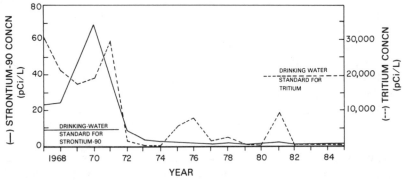

Figure 3. Annual average concentrations are provided of ⁹⁰Sr and tritiated water measured in continuously-collected water samples from Cattaraugus Creek during the period 1967 to 1981 and in grab samples from 1982 to 1985. For easier presentation, the annual average concentration value for each year is plotted as a single point at the marker for that year; the points are joined by straight lines. Standards promulgated by the U.S. Environmental Protection Agency for the National Safe Drinking Water Act are provided for reference. Concentration values through 1971 include reprocessing effluents, as well as precipitation pumped from active trenches. Values for 1972 through 1974 represent releases from decontamination of the reprocessing plant along with those from precipitation pumped from active burial trenches. From 1975 on, concentration value include effluents from plant decontamination and releases from pumping of closed trenches.

operate a low-level liquid-waste treatment facility which reduced the concentrations of all measured radionuclides except tritium. From 1972 to now, concentrations of all radionuclides in Cattaraugus Creek have been well below U. S. Environmental Protection Agency (EPA) and DOH standards for drinking water (the standards for ⁹⁰Sr and tritium are also indicated on Figure 3). Tritium and ⁹⁰Sr are the most important of the radionuclides in the liquid effluents from either the reprocessing plant or the waste burial site, and are the two specifically limited in EPA's 40 CFR Part 141 and the N. Y. State Health Code, 10 NYCRR Part 5. The slightly elevated concentrations of tritium which occurred during 1975, 1976, 1977, 1978, and 1981 were for the most part due to pumping water out of the burial trenches; some tritium and ⁹⁰Sr has also been released from an ongoing decontamination program at the reprocessing plant, and some comes from fallout.

Some tritium and ¹⁴C leaves the burial site as gases formed in the biodegradation of buried organic wastes.[10,11,12] Based on measurements made from 1976 through 1979, approximately 90% of the ^{14}C will escape as gas through the trench covers over a period of ten to fifty years.[12] Also, it appears as if approximately one-third of the tritium will vent through the trench covers.[12] More accurate estimates of time and quantity cannot be made, because the studies were only conducted over a short period in the life of the biodegradable wastes; another evaluation appears appropriate at this time.

One other important radionuclide, radon-222 (^{222}Rn), is produced from the decay of buried radium-226 (^{226}Ra) and is swept along by the gases formed through biodegradation. Normally ^{222}Rn would not create sufficient partial pressure to escape the trench covers. However, the large volumes of methane and carbon dioxide produced from biodegradation of organic wastes provides sufficient pressure to sweep the radon into the atmosphere.

The data described previously, and samples collected by DEC from various wells on and around the site (reported in DEC Annual Reports on Environmental Radiation) show that radionuclides from the burial site are not contaminating the groundwater system. Only by escaping through the trench covers to surface water and to the atmosphere can the radionuclides impact on people.

WHAT ARE THE REGULATORY AND HEALTH IMPLICATIONS?

Monitoring of onsite surface water supplies in Lakes Erie and Ontario has been ongoing since 1965. The results for these sites have been reported annually by DOH (prior to 1969) and by DEC (since 1969) along with the results from the continuous sampler at Springville Dam. Radionuclides from the burial site, even during periods when the trenches were pumped, cannot be identified above ambient levels at any of the public water supplies downstream. If we consider as an example Angola, the nearest public water supply at 35 miles downstream from the site, approximately 8000 Ci of tritium have been discharged since 1974 due to pumping of water from the trenches into Buttermilk and Cattaraugus Creeks to be diluted as it flowed downstream into Lake Erie. During that same time more than 110,000 Ci of tritium flowed through Lake Erie—the result of weapons test debris and natural production by cosmic rays. Burial site tritium represents only 7% of the tritium at Angola, and lesser amounts at other water supplies further downstream. Such a small portion of the total tritium inventory cannot be distinguished by any measurement technique, because the periodic variations in the ambient concentration of tritium are far greater than that caused by the burial-site effluents. Even during trench pumping, tritium concentrations in Cattaraugus Creek at Gowanda, only 15 miles below Springville Dam, became indistinguishable from the concentration in rainfall. Other radionuclides are even more difficult to find, because exchange into stream sediments greatly decreases the already small concentrations even more than that accomplished by dilution alone. During the pumping of trenches, none of the applicable DOH regulations was ever exceeded; in fact, radionuclide concentrations did not exceed even 1% of the applicable DOH regulations (10 NYCRR Part 16 or 10 NYCRR Part 5).

Another route by which waterborne radionuclides might impact on people is from their uptake by fish and subsequent ingestion of the fish. The USEPA has performed studies on dose and risk to man from reprocessing-plant effluents at West Valley.[18-20] Using the EPA caculation methods and ingestion data, the low concentrations and quantities of radionuclides in burial-site effluents produce trivially small doses.[2]

Dose commitments and associated risks estimated from analysis of the radionuclide content of water and fish samples are summarized in Table 1. Since 1975, nearly 11-million liters of treated leachate have been pumped into the local streams. The dose to the public water supply customer has been no greater than 0.02 mrem/y, a value which is 0.5% of the 4 mrem/y permitted by EPA (40 CFR Part 141) and DOH (10 NYCRR Part 5) regulations for drinking water. The dose to a fisherman who catches all of his fish in Buttermilk Creek and eats the bones as well as the flesh would receive at most 0.04 mrem/y, a value which is less than 0.2% of the 25 mrem/y permitted by NRC's regulations for shallow-land burial, 10 CFR Part 61, and by EPA standards for fuel-cycle facilities, 40 CFR Part 190. The consequences of water treatment before release (100-fold reduction), obtaining most of the catch from Cattaraugus Creek (10-fold reduction), and boning the fish before eating (at least 100-fold reduction) provide a more realistic estimate of the fisherman's dose and risk (approximately a hundred-thousandth of the values in Table 1).

Not only are the annual doses well within presently prescribed limits, they are well within the value of 1 mrem/y which NRC has proposed as a de minimis level,[21] which EPA is considering as a level "below regulatory concern," and which the National Council on Radiation Protection and Measurements has equated to a "negligible risk."

Despite poor management of the wastes sent to burial at West Valley and despite poor management of the site by NFS, we find that the doses from waterborne effluents from the site are not only acceptable within existing regulations, but are also below further regulatory concern even when applying the most stringent criteria proposed by federal agencies.

The dose due to gases emanating from the site are poorly defined,

Table 1. Maximum Values of Dose and Health Effects from Annual
Release of One Million Gallons of Untreated Trench Water

| | Dose Commitment | |
	Individual (mrem/y)	Population (person-rem)
PWS customer[a]	$< 2 \times 1^{-2}$	< 50
Fisherman[b]	$< 4 \times 10^{-2}$[c]	$< 1 \times 10^{-2}$

| | Risk (Mortality/y) | |
	Individual	Population
PWS customer[a]	$< 2 \times 10^{-9}$	$< 5 \times 10^{-3}$
Fisherman[b]	$< 4 \times 10^{-9}$[c]	$< 1 \times 10^{-6}$

[a]The largest population group (approximately 2 million public water supply
customers).
[b]The maximum individual, a fisherman who eats the entire fish.
[c]Since no one eats the entire fish, a more realistic value of dose to the fisherman is
$< 4 \times 10^{-7}$ mrem/y and of risk is $< 4 \times 10^{-14}$ (see text).

however. Estimates of dose rates from gases transported to the fenceline
of the burial site result in a millionfold range of values; the smallest
values are again trivial, the largest exceed current regulations. Better
estimates for the gaseous effluents cannot be made, because further field
work was stopped prior to 1981 due to transfer of site responsibility from
NFS to the New York State Energy Research and Development Authority (NYSERDA) and subsequently from NYSERDA to DOE. Continuing the gas-effluent studies seems to be necessary for maintenance of the
existing site as well as for planning for future burial sites anywhere in the
humid eastern United States. The fact that no one lives, or has lived, at
the fenceline of the West Valley burial site, leads us to conclude that the
public has never received an annual radiation dose from gaseous effluents which exceeds, or is even an appreciable fraction of, the dose values
estimated for the surface water pathways.

ARE THE DOSES FROM WEST VALLEY REALLY TRIVIAL?

While comparing doses to regulatory limits may have meaning to radiation specialists, such small doses are difficult for most people to comprehend. For example, how might a judge, a lawyer, a member of a jury,
or a legislator decide what dose is "trivial," "below regulatory concern,"
or "de minimis?" One of the more interesting perspectives—the one
which leads to the theme of this chaper—is that of comparing radiation
doses received from the West Valley burial site to those from sleeping
with another person. Estimates of the annual dose received from a bed-

partner range from 3 mrem/y[22] to 0.1 mrem/y[23]. The 30-fold difference between the two estimates depends on a variety of assumptions, but a large factor is how closely the two people are assumed to sleep. It turns out that the difference is nearly the same as that from two people sleeping in an ordinary double bed as opposed to their sleeping in a king-size bed. The dose from sleeping in twin beds in the same room falls below those for a king-size bed and is highly dependent on how the beds are arranged; the dose from sleeping in separated twin beds might be as low as 0.05 mrem/y.

Thus, we can see that radiation dose commitments to the public from releasing one million gallons per year of untreated burial-site effluents are approximately 100-fold less than those you might receive from sharing a double bed with someone, and less than one-half to one-tenth those from sharing any large bed. Only the doses from sleeping in separated twin beds can be considered as trivial as those caused by effluents from the West Valley burial site.

CAN OR SHOULD MORE BE DONE TO REDUCE EFFLUENTS?

If effluents from the commercial burial site at West Valley have caused only a trivial change in the radiation exposures normally experienced by anyone of the public, can or should any more be done? From the regulatory standpoint, no more control is warranted; the effluents are well "below regulatory concern" as defined by any federal agency. However, public fear of radiation, regardless of how small the dose, at first seems to warrant more control over burial-site effluents. Several options are available from work being performed in other countries.

Incineration of biodegradable wastes appears to offer one mechanism for better control of effluents by providing for better management of the burial site through the reduction of void space and of gas buildup. Several countries, including Sweden, Belgium, and Germany, are developing technologies which include the incineration of all combustible wastes, although the requirement of incineration seems to have evolved more from controlling fire hazards during interim storage or burial operations than from post-closure requirements. France does not require incineration of solid waste received for disposal at its site. Incinerator designs vary in complexity from hand-loaded systems (as in Germany) to the fully automated and heavily shielded unit in Belgium. The latter produces a high quality, insoluble ceramic waste product which further enhances management of the burial sites. However, there seem to be no dose savings from the use of incinerators. Most of the radioactivity released as gaseous combustion effluents will be tritium and ^{14}C; but, tritium and ^{14}C are the only radionuclides at the West Valley burial site of which a large fraction of the inventory will be released, also mostly as gases. Incineration of waste will only transfer the release point from the burial site to the incinerator, where releases will occur immediately rather than over several years. Whether better management of the disposal site should take precedence over the increased fiscal costs for incinerator construction and operation is not clear at this time. Also, total dose will be greater, because both incinerator operators and burial site workers will receive radiation doses when wastes are incinerated.

The use of other options must also be weighed on the basis of increased fiscal costs. The disposal site being built under the Baltic Sea by Sweden is expensive.[24] Disposal in mined granite or salt formations, as planned in Germany[25] and Switzerland[26] is also likely to be expensive. The plan to dig deep into clay in Belgium[27] also appears to increase disposal costs over those for shallow-land burial. Interestingly, one of the disposal systems proposed for use in Canada will place the wastes into near-surface pits dug into sand, using a clay/sand mixture and concrete as barriers against water infiltration — the sandy soil will provide assurance that trenches will not fill and overflow.[28]

Unfortunately, except for the system now used in France, none of the systems being developed elsewhere has any operating history, so neither fiscal costs nor worker exposures can be compared effectively to those for shallow-land burial. The French system does have 10 years of operating history, so it can provide some insight on costs and worker exposures. Unfortunately, 10 years is too short a period to provide assurance that effluents will never develop from use of the French system.

The French claim their costs are competitive with those for shallow-land burial.[29,30] However, it is not clear that their cost accounting is complete, so the cost of using the "French system" in the United States may prove to be greater than those published so far. (Although Lavie refers to 13 years of experience as of 1984[30], only 8 of those years[29] are with what is now popularly called "the French system."

More important than fiscal cost is the issue of worker exposures. Radiation doses in 1981 to workers at the La Manche disposal site in France were 1000 mrem/y to the average individual and 2000 mrem/y to the maximum individual.[31] Radiation doses at the Barnwell, South Carolina, site averaged a little less than 400 mrem/y during the same time and have decreased to 245 mrem/y in 1985.[32] Merz[25] estimates the average dose to the workers at the German granite facility will be 3000 mrem/y, a value which should be representative of doses to workers at the mined facilities in Sweden and Switzerland as well. In a recent technical position statement[33], the NRC stated, "Worker exposure and safe operations should obviously be a factor in developing designs" (as alternatives to shallow-land burial for low-level radioactive waste disposal). It is too soon to know how that statement will be translated into licensing action.

Not only are the individual and average worker doses significantly greater at La Manche than at Barnwell, but the worker doses at La Manche result in calculated individual risks which are 100,000 times greater than the worst values for PWS customers and at least 50,000 times those for a fisherman downstream from West Valley. As a result, there is no overall dose saving from adopting the French system; the population dose commitments are of the same order as those for the highest doses which might result at West Valley. For example, from Table 1, the 2-million PWS customers on Lakes Erie and Ontario would experience a 70-year dose commitment of less than 50 person/rem, while the 30 to 70 workers required to operate a burial site the size of La Manche would also experience approximately 50 person-rem as their dose commitment. Thus, the statistically calculable number of deaths (0.005 per year) will be approximately the same whether shallow-land burial or the French system is used. The main difference is that the West Valley estimates are for a hypothetical worst case condition, whereas the worker doses in France are real and continuing. The statistically-calculable radiation deaths for the mined facilities in Germany, Sweden, and Switzerland are threefold greater that those at West Valley or La Manche.

If the number of calculated deaths is the same or more for alternate systems, then the large difference between individual worker doses is the determining factor. It should be of concern to health officials and may be particularly important to lawyers involved with tort actions related to radiation injuries. In response to a 1982 amendment (P.L. 97–414) to the Federal Food, Drug and Cosmetic Act, the National Institutes of Health developed radioepidemiological tables[34] from which experts can make reasonable estimates of the probability of causation of a cancer developed by an individual. If the annual dose to an individual worker from use of the French system is 100,000 times that to any member of the public, the causation tables generally will show a proportionately greater probability of causation for any cancer in the worker population and may be used in any tort action against an owner/operator of a U. S. disposal site using the French system.

NRC regulations, 10 CFR Part 61, require that burial-site effluents should not cause a dose of more than 25 mrem/y to anyone living offsite, a dose limit approximately 1000-fold greater than the doses from West Valley. The NRC regulations also require that the dose to a maximally-exposed worker not exceed 5000 mrem/y, a limit which has a margin of only 2.5 times the dose to the maximally exposed individual at La Manche. The French system may, therefore, violate the "as low as reasonably achievable"

(ALARA) principle mandated by NRC regulations. The ALARA principle has proven an important legal and technical issue between radiation workers and their employers at many nuclear facilities in the United States.

The high doses to the burial site workers in France must also be considered in relation to the other developing technologies. Because disposal using the French system is performed at land surface[29,30], ease of handling will keep total dose low compared to that for each of the other alternative systems. Burial in more confined spaces, such as those in the systems planned for Sweden, Germany, or Switzerland, is likely to cause either greater worker doses or the distribution of a larger total dose over a greater worker population to keep individual doses within regulatory limits. No matter which is the choice for dose distribution, the number of calculated deaths will increase, as will the likelihood of legal action.

While the public demands may make necessary the use of an alternative technology, it is not clear that any of the developing technologies is the one of choice. All will cost more than shallow-land burial. All are likely to cause greater dose to workers. Even incineration may only result in a transfer of dose from one population to another, while also adding worker exposure at the incinerator. Transferring dose to workers, or to other populations, may provide the cosmetic appearance of providing greater safety at the burial site, but in reality the overall health impact from these alternative technologies appears to be at best the same as, or possibly worse than, that from shallow-land burial.

If the choice is made primarily to assuage public apprehension, then the most economical choice seems to be in selecting the proper site for shallow-land burial, such as sand as at Chalk River, Canada, or a clay-sand mixture as at Barnwell, South Carolina.[16] At each of these locations, the soil underlying the trenches is more permeable than the cover material. Unlike West Valley, any leachate from the waste buried at Chalk River or Barnwell will seep downward into the groundwater system, and will not be visible or easily accessible for analysis. While such sites may appear successful, it is not clear that any real dose savings results when comparing the sites to West Valley. Allowing leachate from a radioactive waste burial site to seep into groundwater only opens new fields for regulatory action or litigation.[35]

SUMMARY

Despite burial of unacceptable waste forms and poor management of the site, annual doses from shallow-land burial at the West Valley site are well within existing and forthcoming standards. Even the maximum annual health risks, less than two in one billion for a public water supply customer and less than four in one billion for a fisherman result in lifetime risks for each which are more than 100-fold below "one in a million," a value commonly considered acceptable.

Decomposition of buried biodegradable wastes will result in the production of large volumes of gas which in turn will create problems with management of a disposal site, even if a system is used similar to that used in France. Since tritium and ^{14}C will escape to the atmosphere regardless of whether or not the wastes are incinerated, management of any disposal site may be improved by incineration prior to burial. It appears that worker and public doses will increase if incineration is used, however.

Several new technologies can be considered to improve management and performance of a burial site. Whether such expensive improvements are necessary depends solely on the need to assuage public fears rather than on technical merit. Use of the "French system" would result in adoption of an alternative disposal technology which has to date caused greater worker doses than those from shallow-land burial; all other technologies using disposal at depth have the potential for causing even greater worker doses than does the French system. The potential radiation doses from West Valley burial site effluents are of the order of those from sleeping in separated twin beds and are approximately 3- to 100-fold less than those from sharing the same bed with someone, so selection of any other currently known technology seems difficult to justify from a purely technical standpoint.

REFERENCES

1. Kelleher, W.J., "Water Problems at the West Valley Site," in *Management of Low-Level Radioactive Waste*. Carter M. W., Moghissi, A.A. and Kahn, B., Eds., 843–851. Pergamon Press, New York, 1979.
2. Matuszek, J.M., Strnisa, F.V., and Baxter, C.F., "Radionuclide Dynamics and Health Implications for the New York Nuclear Service Center's Radioactive Waste Burial Site," in *Management of Radioactive Waste from the Nuclear Fuel Cycle*, Vol. II, 359-373. International Atomic Energy Agency, Vienna (1976).
3. Prudic, D.E. and Randall, A.D., "Groundwater Hydrology and Subsurface Migration of Radioisotopes at a Low-Level Solid Radioactive Waste Disposal Site," in *Management of Low-Level Radioactive Waste*, 853–882. Pergamon Press, New York, 1979.
4. Prudic, D.E., "Permeability of Covers Over Low-Level Radioactive Waste Burial Trenches, West Valley, New York."; U.S. Geological Survey, Water Resources Investigation Report 80–55, 1980.
5. Prudic, D.E. *Ground Water* 1982, *20*, 194.
6. Fakundiny, R.H. *Northeast Environmental Science* 1986, *5*, in press.

7. Lu, A.L. *Health Physics*, 1978, *34*, 39–44.
8. Husain, L., Matuszek, J.M., Hutchinson, J. and Wahlen, M., "Chemical and Radiochemical Character of a Low-Level Radioactive Waste Burial Site," in *Management of Low-Level Radioactive Waste*, 883–900. Pergamon Press, New York, 1979.
9. Matuszek, J.M., Husain, L., Lu, A.H., Davis, J.F., Fakundiny, R.H. and Pferd, J.W., "Application of Radionuclide Pathways Studies to Management of Shallow, Low-Level Radioactive Waste Burial Facilities," in *Management of Low-Level Radioactive Waste*, 901–914. Pergamon Press, New York, 1979.
10. Lu, A.H. and Matuszek, J.M., "Transport Through a Trench Cover of Gaseous Tritiated Compounds from Buried Radioactive Wastes," in *Behaviour of Tritium in the Environment*, 665–670. IAEA, Vienna (1979).
11. Kunz, C., *Nucl. & Chem. Waste Management* 1982, *3*, 185–190.
12. Matuszek, J.M., "Respiration of Gases from Near-Surface Radioactive Waste Burial Trenches," in *Waste Management '83 Symposium*, Vol. I, 423–427. University of Arizona Press, Tucson, 1983.
13. Prudic, D.E., "Core Sampling Beneath Low-level Radioactive-Waste Burial Trenches, West Valley, New York." U.S. Geological Survey, Open File Report 79-1532, 1979.
14. "An Aerial Radiological Survey of the Nuclear Fuels Services Center (NFS) and Surrounding Area, West Valley, New York." EG&G Report, EGG-1183-1782, 1980. Available from NTIS, Springfield, VA.
15. "1971 Report of Environmental Radiation in New York State," 11–40. N.Y. State Department of Environmental Conservation, Albany, NY.
16. Neiheisel, J., "Prediction Parameters of Radionuclide Retention at Low-Level Radioactive Waste Sites," EPA 520/1-83-025. U.S. Environmental Protection Agency, 1983.
17. Scholle, S.R. *Northeastern Environmental Science* 1983, *2*, 8.
18. Schleien, B., "An Estimate of Radiation Doses Received by Individuals Living in the Vicinity of a Nuclear Fuel Reprocessing Plant in 1968," BRH/NERHL 70-1. U.S. Dept. of Health, Education and Welfare, 1970.
19. Martin, J.A., Jr., *Radiological Health Data and Reports* 1973, *14*, 59.
20. Magno, P.J., Kramkowski, R., Reavey, T. and Wozniak, R., "Studies of Ingestion Dose Pathways from the Nuclear Fuel Services Fuel Reprocessing Plant," EPA-520/3-74-001. U.S. Environmental Protection Agency, 1974.
21. Federal Register 1985, *50*, 52014.
22. Teller, E. *WPI Journal* 1974, *78* (1), 9.
23. "Radiation." Westinghouse Electric Corporation, 1979.
24. Nilsson, L.B., "Geologic Disposal of LLW in Sweden," in *Alternative Low-Level Waste Technologies,* Illinois Department of Nuclear Safety (in press), 1986.
25. Merz, E.R., "Geological Disposal of Low-Level Radioactive Waste in the Federal Republic of Germany," in *Alternative Low-Level Waste Technologies,* Illinois Department of Nuclear Safety (in press), 1986.

26. Rometsch, R., "Geological Disposal of Low- and Intermediate-Level Radioactive Wastes," in *Alternative Low-Level Waste Technologies*, Illinois Department of Nuclear Safety (in press), 1986.
27. Dejonghe, P.A.J., "Low and Medium Level Radioactive Waste Management in Belgium," in *Alternative Low-Level Waste Technologies*, Illinois Department of Nuclear Safety (in press), 1986.
28. Charlesworth, D.H., "Federal Activities in the Management of Low-Level Radioactive Wastes in Canada," in *Alternative Low-Level Waste Technologies*, Illinois Department of Nuclear Safety (in press), 1986.
29. Barthoux, A.J. and Hutchinson, C., "The French System," in *Alternative Low-Level Waste Technologies*, Illinois Department of Nuclear Safety (in press), 1986.
30. Lavie, J.M. and Marque, Y., "Stockage en Surface des Dechets Solides de Faible et Moyenne Activite en France: 13 Ans D'Experience Pratique," in *Radioactive Waste Management*. IAEA, Vienna (1984).
31. Sousselier, Y., Pradel, J., Chapuis A.M. and von Kote, F., "Capacite Radiologique et Experience Réelle d'un Site de Stockage en Surface," in *Radioactive Waste Management*. IAEA, Vienna (1984).
32. Pervis, J. (1983) and Whittaker, M. (1986), *Chem Nuclear*. Private communications.
33. Federal Register 1986, *51*, 7806.
34. "Report of the National Institute of Health Ad Hoc Working Group to Develop Radioepidemiological Tables." NIH Publication No. 85-2748, Washington, D.C. 1985.
35. "The Management of Radioactive Waste at the Oak Ridge National Laboratory: A Technical Review" (p. 76). National Academy Press, Washington D.C., 1985.

CHAPTER 14

Living the Past, Facing the Future

Michael E. Burns

It has been almost seven years since Congress passed the Low-Level Radioactive Waste Policy Act of 1980 (LLRWPA) to avert an impending low-level radioactive waste (LLRW) disposal crisis. Five years later, with the effective date of the act imminent, without significant progress having been made, Congress was forced to amend the act and delay its effective date for seven years. At this writing, nineteen months after this extension, it once again seems unlikely that the act will be implemented by its statutory deadline of January 1, 1993.

This chapter argues that this policy has been a failure to date and is unlikely to ever accomplish its goals without extensive modifications. Why do I label the policy a failure? In the seven years since the original act was passed, no new LLRW disposal sites have been sited, and it appears that none will be by the end of the seven-year extension period, despite the carrot-and-stick approach adopted by Congress to encourage state compliance.

MILESTONE COMPLIANCE STANDARDS

Although the Nuclear Regulatory Commission (NRC) sets the milestone compliance standards for surcharge rebates, the sited compacts may (and have) set their own standards for continued access to their disposal facilities. The first milestone is already past. The second milestone date is coming soon. By January 1, 1988, nonsited regions must designate a host state and unallied states must submit a detailed siting plan for a disposal facility. Noncompliance results in an automatic doubling of the existing surcharges for the first six-month period and a quadrupling of them for the following six months. States not in compli-

Low-Level Radioactive Waste Regulation: Science, Politics, and Fear, Michael E. Burns, Ed., © 1988 Lewis Publishers, Inc., Chelsea, Michigan—Printed in USA.

ance a year after the deadline face an automatic loss of their access to the disposal sites.

Some states have been acting as if they expected the sited regions to be more flexible in their treatment of regions or states showing "good faith" in their siting efforts, regardless of the outcome. South Carolina's Governor Carroll Campbell dashed these hopes recently when he said, "[South Carolina's state law and the Southeast Compact] enjoy wide support among citizens and state officials. [I] support these policies, see no significant movement away from them, and foresee no change to the laws upon which these policies are based."[1] No one will admit to it, but some states were expecting a more lenient attitude and perhaps did not take the threat of the loss of access as seriously as they should have.

WHERE DO WE STAND AT THIS POINT?

Regional Compacts

Congress approved seven regional compacts that encompassed 37 states with passage of the Low-Level Radioactive Waste Policy Act Amendments of 1985 (LLRWPAA). The 13 remaining states, Puerto Rico and the District of Columbia that are not members of compacts are referred to as unallied states. The three compacts with existing disposal sites are known as "sited" regions (compacts), the other four as nonsited regions.

Sited Compacts

The Northwest Compact will continue to use the existing disposal site at Hanford, Washington. In return for continuing as the compact's host state, Washington will receive guaranteed access to chemical waste disposal sites in Oregon.

The Rocky Mountain Compact has selected Colorado as its next host state. One Colorado town in an economically depressed area actually volunteered to host the disposal facility in order to generate jobs. It was not selected, however, because it did not meet all of the compacts' selection criteria. Two potential sites in Colorado are currently under consideration.

The Southeast Compact selected North Carolina as their next host state. North Carolina wasn't pleased with this "honor" and that compact almost fell apart over the selection. North Carolina argued that the selection process was flawed and threatened to withdraw from the compact if the decision was not reconsidered. Bills that would have with-drawn the state from the compact were introduced before the state legis-lature. Although the host state selection was reconsidered, North Carolina remained the number one candidate for the site. It now appears that the state will remain in the compact, though siting a facility there will undoubtedly encounter strong citizen opposition.[2]

Nonsited Compacts

It appears now that none of the nonsited compacts are likely to com-plete the tasks of selecting a host state, finding a site location, selecting a contractor, preparing and submitting an NRC site application, and con-structing a disposal facility by the 1993 deadline. Some of them have not even selected a host state yet, let alone a host site.

The Northeast Compact, for instance, continues to encounter serious difficulties in establishing a site. Originally envisioned as a coalition of New England and Middle Atlantic states, the compact consists only of Connecticut and New Jersey. The northeastern United States contains some of the nation's largest LLRW-generating states, including Massa-chusetts, New York, and Pennsylvania. Other states were reluctant to form compacts with them for fear that they would be stuck disposing of the larger generator's wastes. To complicate the situation further, until recently, Massachusetts's law required the approval of a statewide refer-endum prior to siting a LLRW disposal facility in the state. Though the Massachusetts Supreme Court has since ruled this law unconstitutional, its existence made an agreement with any other state unlikely.

When approved by Congress, the Northeast Compact consisted of four states: Connecticut, Delaware, Maryland, and New Jersey. Later, Maryland and Delaware decided to withdraw from the compact and form a compact with Pennsylvania and West Virginia (the Appalachian Compact). In this case, the word "withdraw" is something of a mis-nomer. In fact, since neither state ever paid their entry fee to join the Northeast Compact, neither was ever a legal member of it.

Why did Maryland and Delaware devote the legislative and adminis-trative time to string the Northeast Compact along if they were not going to join it? Undoubtedly, they wanted to keep their options open while looking around for a better deal. The idea of a compact with Pennsylva-nia and West Virginia had been floating around for some time, and Pennsylvania, the largest generator of the four, had already agreed to be the host state if the compact was formed. By allying themselves with Pennsylvania and West Virginia, Maryland and Delaware avoided the politically and socially difficult task of siting a disposal facility within their borders. On the other hand, New Jersey and Connecticut were left holding the bag since one of them will now face the burden of siting a facility within their borders.

The withdrawal of the two states brings up an interesting question. The Northeast Compact is a legal contract between its member states. One of its provisions is that its Compact Commission (its legal entity) could not be formed until *three* eligible states ratified the compact. Since only Connecticut and New Jersey ratified the compact, the compact's legal status could be challenged in court. This situation begs the question, Can (or would) Congress grant its consent to a compact that was not legally constituted?

When I asked an official of the Northeast Compact what the compact's position was on the withdrawal of Maryland and Delaware, I was told: "The [compact's] Commissioners have taken no position on this matter. They feel it is up to Congress to resolve this issue." At the same time, congressional staffers told me that Congress had no intention of entering a fray between states and that it was up to the states or the courts to settle any disputes that arise between them. Personally, I think that Connecticut and New Jersey are living in a pipe dream if they think that Congress will solve this problem for them.

Other nonsited compacts have made greater progress toward the establishment of disposal sites. For instance, the Central Midwest Compact—Illinois (host state) and Kentucky—has completed the preliminary phases of its site selection process and issued its request for proposals from site contractors.

Other compacts have narrowed their search for a host state to several member states and are looking for a volunteer among them. Not surprisingly, volunteers have been slow to step forward.[1]

North Carolina is not the only state to threaten withdrawal from a compact after being designated a host state. Both Kansas and Michigan made the same threat when named host state for their respective compacts. Michigan officials began to express their displeasure over their selection as soon as the choice was made. State officials claimed the right to "accept or reject" host state status and announced that they would defer this decision until August 1987. Among other concerns cited by Michigan were "serious deficiencies" in the national regional compact scheme. One specific concern was that there were no guarantees that the host state would be able to recover the full costs of establishing a facility.[3]

Just prior to their selection, Michigan officials wrote to other compact members seeking their support for a move to have Congress reexamine the LLRWPAA and consider replacement legislation. In an interview, Compact Commissioner David Hales of Michigan was quoted as saying,

"If there is *any chance* that there is going to be endangerment of public health, safety, or the environment in the state of Michigan, then Michigan will not build that facility."[4] (italics mine) Isn't it odd that Michigan waited until the month before being designated the regional host state to express their concern about the viability of this national policy? Did it really take almost seven years for Michigan's governor to decide that this law is "seriously flawed?" I find Michigan's concerns hard to accept at face value. This sounds to me like a sophisticated version of a child complaining, "If I have to be *it*, then I'll take my ball and quit the game."

These states are shouting, "Not In My Backyard" (NIMBY). Put it someplace else, anyplace but here. The NIMBY cry is often hidden behind technical objections (e.g., the selection methodology was flawed, the wrong waste volume figures were used) that, oddly enough, only seem to come up after the host state is selected, even though the compact members agreed on the selection criteria long before the actual choice was made. It seems that agreements to share a regional disposal site are easier to reach than are agreements as to *where* the site should be.

Proposed Compacts

Two compacts have been proposed in addition to the seven congressionally approved compacts. They are the Appalachian Compact consisting of Pennsylvania (host state), Delaware, Maryland, and West Virginia, and the Southwestern Compact of California (host state), Arizona, North Dakota, and South Dakota. The Appalachian Compact will probably be considered by Congress in the fall of 1987; the Southwestern Compact has just been submitted to Congress.

The Southwestern Compact has just been enacted and signed into law in California. Since California has agreed to be the compact's host state for at least the first 30 years of the agreement, the other states are expected to quickly ratify the agreement. The agreement allows additional states to join the compact if all its members agree. Some unallied states have already made inquiries about joining.[5]

Unallied Entities

The remaining unallied states face increasing pressure as they approach new milestone dates. Among them are such major LLRW generators as Texas, New York, and Massachusetts.

Texas. Many people feel that Texas will be the first unallied state to site a facility. In February 1987, the Texas LLRW authority completed a survey of state-owned land and chose a site located about 50 miles southeast of El Paso. Unfortunately, the land commissioner refused to transfer title to the land to the authority unless ordered to do so by the state legislature. A bill directing the transfer and banning the use of shallow land burial (SLB) as a disposal technology was passed and signed into law in July 1987. With passage of this legislation, the LLRW authority is expected to go to court and request dismissal of a temporary injunction that forbids further work at that site.[6]

This delay may mean that Texas will not meet the 1988 milestone. If not, its generators will begin paying surcharges on wastes shipped to existing disposal sites. Based on the current rate of generation, these surcharges will amount to $35 million by 1993.[6] (This assumes that none of the sites denies Texas access. I doubt that this will happen, however, for if any state has demonstrated "good faith" in their siting efforts, Texas has.)

New York. New York decided to site its own facility after failing to reach a compact agreement with any other state. Like Massachusetts, New York generates too much LLRW for another state to be interested in a compact unless New York agrees in advance to be the host state. In addition, the failure of the West Valley disposal site has clearly had a negative influence on public attitudes toward LLRW disposal in the state. The New York siting bill specifically prohibits the further use of the West Valley site and bans SLB as a disposal method.

Recently, the New York State Department of Environmental Conservation (NYDEC) recommended the establishment of a $9 million incentive package and a strong local host commission to encourage community acceptance of a disposal site. The accompanying report listed some of the negative impacts a disposal site could have on a local community and recommended specific "offsets" that could be used to mitigate these impacts. Among them are annual payments to the host community, compensation for decreased property values, and the development of a "Host Area LLRW Commission" with the power to recommend (not force) the temporary or permanent closure of the facility. This commission would also be responsible for ensuring citizen participation by convening public meetings at least monthly for the first year after the site's operating license is granted.[7] It remains to be seen whether or not the state legislature will go along with these recommendations, and, if so, how successful they will be in application.[8]

Other compacts are considering similar programs. For instance, Illinois's siting law provides funding for local citizens' groups to hire independent experts to review siting proposals. I find these proposals mildly encouraging because they suggest that state regulators are studying the research that Bord and others have done on the public's attitudes about locally unpopular facilities.[9] Lawmakers, regulatory agencies, and commercial entities are going to have to consider the sociological and psychological impacts of these facilities if unpopular facilities such as radioactive or chemical waste disposal facilities are to be sited successfully.

Massachusetts. Although the statute requiring the approval of a statewide referendum before a LLRW disposal facility could be sited was struck down by the state's supreme court, Massachusetts still faces major problems in siting a disposal facility because of the large volume of LLRW it generates. So far, it has been unable to reach an agreement with any other state.

A siting bill was introduced before the state legislature in 1986. As of this writing, it has not yet passed the Massachusetts House of Delegates, though some observers feel that it has a good chance of passage this year. The bill has aroused a great deal of debate, suggesting that even if it passes, substantial political and public opposition can be expected. Governor Dukakis has been a strong supporter of the bill and is expected to sign it into law if it passes.

Even if this bill were signed into law and funds to implement it were available immediately, it is unlikely (to put it kindly) that the state could site, permit, and construct a facility before the January 1993 deadline. (The NRC's official estimate of the time needed to site, permit, and construct a disposal facility is five to seven years but this estimate does not include delays caused by civil litigation, which could lengthen the process by years.) Can anyone doubt that civil litigation will occur over an issue that arouses as much public controversy as radioactive waste disposal?

Other Unallied States. The remainder of the unallied states generate relatively small volumes of LLRW. I suspect that most of them will eventually join an existing compact or form one or more among themselves. None have announced plans to site their own facilities.

Some are attempting to negotiate contracts for waste disposal with the sited compacts. This idea is attractive to the sited compacts from an economic standpoint. The additional waste volume their sites would handle is insignificant and they could (and undoubtedly would) charge a hefty premium for this service. On the other hand, contract disposal of "foreign" wastes could turn out to be a political disaster. The LLRWPA of 1980 was enacted largely because of the objections of the three states with disposal sites to being the nation's LLRW dumping grounds. Under a contract arrangement they would still be the dumping grounds — but they would be paid more for the service. The additional revenues generated by contract disposal are unlikely to be enough to make these arrangements acceptable to the citizens of those states.

Mixed Wastes

Mixed wastes are a subset of LLRW that contain both radioactive and hazardous chemical wastes. Although they represent only a small portion (an estimated 1–3%)[10] of the total LLRW generated each year, they are currently the most difficult portion to dispose of because there is no disposal option available for some of them.

Why are mixed wastes so hard to dispose of? First, they fall under the authority of two regulatory agencies. The NRC regulates radioactive waste, while the Environmental Protection Agency (EPA) is responsible for chemical wastes. Furthermore, the disposal requirements for the two types of waste differ sharply; for instance, chemical wastes that are buried must be placed in a trench with an impervious liner, but liners are expressly forbidden for SLB of LLRW.

To make matters worse, until recently, neither agency would yield to the other's requirements, both arguing that they lacked statutory authority to yield any regulatory authority over the hazard component they regulated. Congress made it clear in 1985, before passing the LLRW-PAA, that they would provide the required statutory authority to either agency if they would agree on what was needed. Both insisted that they did not require congressional relief to resolve the issue. They remained "confident" that an agreement would be ironed out.

Finally, in January 1987 the agencies issued a joint guidance document on the definition and identification of mixed wastes. After eighteen months of joint effort, generators now knew how to identify what they could not get rid of. Adding insult to injury, the document essentially defines mixed wastes by combining the definitions of radioactive and chemical wastes found in NRC and EPA regulations. It also provides some typical examples of mixed wastes.[11]

In April 1987, the agencies issued a joint guidance document containing a set of 11 siting guidelines for LLRW facilities that will handle mixed wastes. Significantly, this document provides guidance rather than regulations since each agency still plans to issue its own regulations on site suitability, handling practices, recordkeeping, and so on. Presumably, as a result of two years of their joint efforts, the regulatory inconsistencies will have been worked out.[12]

Joint regulation of mixed wastes by EPA and NRC could have important legal ramifications. Chemical wastes are treated differently than radwastes under existing law. The Comprehensive Environmental

Response, Compensation and Liability Act of 1980 (CERCLA), more commonly known as Superfund, makes the generator of chemical wastes perpetually liable for their safe and permanent disposal. Thus mixed wastes may come back to haunt generators, as they have at Maxey Flats.

Vanishing Disposal Options

In November 1985, the Hanford site stopped accepting mixed wastes for disposal when state authorities told the site operator that it would have to comply with both the hazardous and radioactive waste regulations. At best this would have been an onerous task, at worst an impossible one. So the site operators stopped accepting mixed wastes for disposal. As a result, there is no legal disposal option for some types of mixed wastes.

Generators are being forced to store these wastes until another disposal option becomes available. Some large generators are already storing thousands of gallons. They may be stuck with them for a long time because none of the existing sites are anxious (or likely) to get back into the chemical waste side of the business. Specific siting guidelines for mixed wastes are not even expected from EPA until sometime in 1988. The delay in the promulgation of siting criteria, concern over the possibility of fostering future Superfund sites, and the relatively small proportion of LLRW that mixed wastes represent suggests that the disposal of these materials will continue to be a difficult problem for the foreseeable future.

Mixed Wastes and Superfund

The generator's perpetual liability under Superfund is a key factor that is making mixed wastes difficult to dispose of. The commercial LLRW disposal site at Maxey Flats, Kentucky operated from 1963 to 1977. About 4.8 million cubic feet of radwastes containing over 2.5 million curies of radioactivity were disposed of in trenches and pits on the site. Much of the waste consisted of solid materials such as contaminated paper, trash, clothing, laboratory equipment, and animal carcasses. A much smaller portion consisted of contaminated organic chemical solvents such as benzene, toluene, and xylene.

In the last few years, some of these solvents have been detected in monitoring wells outside of the active portion of the site indicating that they have migrated from the burial trenches. As a result, the site has been placed on the National Priority List as a Superfund site. To date, the

EPA has identified over 800 potentially responsible parties, i.e., waste generators who sent materials to the facility for disposal and will be required to share the costs of contamination assessment and cleanup of the facility.[13]

Maxey Flats is not the only Superfund site that involves mixed wastes. New Jersey has removed some 15,000 drums of radium-contaminated soil from around houses built on what was apparently a dump site for a radium dial factory. The drums have been stored on state land for several years. Three years ago, local activists got a court to order their removal because they represented a "traffic hazard."[14] The state then made plans to ship them by railroad to the disposal site in Beatty, Nevada. But Nevada authorities immediately brought a suit to block the plan because the railroad's plans to unload the material at a siding near Las Vegas might represent a public health threat.

Then the other two commercial LLRW sites refused to accept the material *because the level of radium contamination in the soil was too low to warrant its disposal in a LLRW site.* First it is a threat to public health, then it is not radioactive enough to dispose of. Recently, the U.S. Department of the Interior rejected a New Jersey plan to store the drums in a federal nature preserve. So for the moment, the drums remain stored on state land over the strenuous objections of the local citizens.

Now for the story's ironic ending. The state's latest plan is to mix the soil with additional radioisotopes to make it radioactive enough to be accepted at a disposal site. You say that increasing the radioactivity of the soil doesn't make any sense at all? It does under the LLRWPAA because the sites are obligated to accept LLRW until 1993 (subject to certain site-specific volume restrictions) as long as the generating state has not been refused access and the packaging and paperwork are in order.

LLRW Volumes Are Decreasing

Our annual LLRW volume has decreased sharply since passage of the LLRWPA. In 1979, 2.6 million cubic feet of LLRW was shipped to disposal sites. In 1986, only about 1.8 million cubic feet (roughly 400,000 33-gallon household trash cans) of LLRW was shipped. And through the end of April 1987, about 545,000 cubic feet has been shipped. At this rate, the 1987 waste volume will total only about 1.6 million cubic feet. This represents a 43% decrease in waste volume since the 1980 statute was enacted.

The American College of Nuclear Physicians reports that LLRW generated by medical sources decreased 70% between 1979 and 1984 and that an additional 20 to 30% decrease in medical LLRW volume can be accomplished.[15]

These volume decreases are largely due to increased efforts by generators — especially large ones — to reduce their waste volumes for practical and economic reasons. The surcharges already in effect for many states and the ever increasing costs of disposal have provided a powerful economic incentive for generators to minimize waste volumes by modifying work practices and waste handling procedures. Some of the waste minimization techniques being used by generators include: (1) segregation of short-lived radioisotopes followed by onsite storage for decay, (2) increased use of supercompactors to reduce waste volumes, (3) recycling radioisotopes, and (4) minimizing the amounts of radioisotopes used for procedures (and consequently minimizing waste). Furthermore, the 1979 figures included an unknown quantity of mixed wastes which are no longer being accepted for disposal. Some of the reduction may also be due to generators who are storing wastes onsite to avoid surcharges until their region (or state) has access to its own disposal facility. So this waste may "reappear" in the statistics for some future year.

It is too early to tell what effect these practices or the lessened use of radioisotopes will have on scientific research. The American College of Nuclear Physicians has stated, "The cost of LLRW disposal and fear of adverse public reaction is causing universities and other institutions to pressure researchers to diminish use of radioactive materials. This will impede research, increase the cost, and slow development of new pharmaceutical materials."[16]

WHERE ARE WE HEADING?

If we are not careful, we are going to paint ourselves into a corner on this issue. It's naive to believe that the public's adamant opposition to disposal sites will be eliminated by congressional fiat. It should be obvious to anyone who has examined this issue that there is no possibility of establishing the required new disposal sites by the 1993 deadline. Must we once again, wait until the last day of a legislative session to modify this policy?

Congress is likely to stick with this concept until forced to abandon it. They'll argue that the states wanted this responsibility, it was given to them, and now the states are not living up to their part of the bargain. This argument ignores the fact that Congress took the easy way out back in 1979, even though the necessity for a new national policy on radwaste disposal had been evident for years. Instead of developing a rational approach to the problem, they adopted a quick and dirty, politically acceptable solution by arbitrarily segregating LLRW, and then dumping responsibility for its disposal into the laps of the states. It should have been obvious that their "solution" was neither workable nor socially acceptable. Now, the only way out of this corner is for Congress to readdress the issue and develop a new approach to the problem.

HOW DID WE GET INTO THIS PREDICAMENT?

The Real Problem Is Waste in General — Not Radioactive Waste

It should come as no surprise that we have trouble disposing of radioactive wastes when you consider that we have never developed a workable system to dispose of the trash we generate daily. Siting landfills for municipal refuse has been difficult for years. When it is time to site a new landfill, everyone yells NIMBY. If we cannot handle ordinary household garbage, how can we handle radioactive or chemical wastes?

Garbage disposal became a subject of national attention recently when a barge loaded with municipal refuse became a modern version of the *Flying Dutchman* and a humorous fixture of daily news reports.

The odyssey began on March 22, 1987 when the tug *Break of Dawn* weighed anchor in New York Harbor and sailed for North Carolina. It towed a barge loaded with 3100 tons of baled municipal waste from the town of Islip, New York. The entrepreneurial owner of the barge planned to use the waste to start a waste-to-energy project at a privately owned landfill in Morehead City, North Carolina. But when the barge reached North Carolina three days later, state authorities refused to allow its cargo ashore because New York authorities had not sampled its cargo before loading and thus could not certify that it was free of hazardous wastes.

The barge was then towed to Louisiana where state authorities inspected its cargo and obtained samples of its leachate before turning it away because the refuse was baled, making identification impossible, and it contained medical wastes which could be infectious.

The barge then sailed to Mexico, but Mexican patrol boats and reconnaissance planes intercepted it and prevented it from entering their waters. Next, on to the tiny Central American country of Belize which also cried NIMBY.

The tug dropped anchor off Key West, Florida on May 2 while its owner thought over his dilemma. Five days later, while it was still at anchor, the Bahamian government weighed in with a prophylactic cry of NIMBY by asking the U.S. embassy to notify the barge owner that they would not accept the cargo either. On May 12, the Commissioner of the

New York Department of Health (NYDOH) announced that an agreement had been reached between Islip and his department that would allow the town to reopen its landfill and dispose of the refuse.[17]

On May 16, the barge and its contents arrived back where it started its 6100 mile journey.[18] But the NYDOH's agreement with Islip did not impress officials of the NYDEC, who stepped in and refused to allow the barge to dock because the cargo might contain "hazardous materials." The battle then entered the court system, where, on May 20, a New York State Supreme Court Justice ordered the NYDEC to develop a plan to dispose of the garbage within six days.[19]

Six days passed (in fact, six weeks went by), while the barge rocked at anchor and its cargo festered. On July 11, the Commissioner of the NYDEC announced an agreement between his office and the New York City Department of Sanitation that would allow the waste to be incinerated at the city-owned facility in Brooklyn. The Department of Sanitation estimated that it would take ten days to burn the garbage and truck the resulting 400 tons of ash to the landfill at Islip.[20]

Once again this proved to be an overly optimistic timetable, however, because the local borough president and the New York Public Interest Research Group immediately sought and obtained a court order blocking the plan. Both parties claimed that the barge contained hazardous chemical and infectious medical wastes. Two days later, unionized city workers at the incinerator refused to handle the barge's cargo, because they were afraid they might contract acquired immunodeficiency syndrome (AIDS).[21]

What is the ending to the story? Well, after a 6000-mile journey that lasted 156 days, and being rejected by six states and three foreign countries, the bargeload of refuse finally docked at the Brooklyn incinerator on August 24, 1987. And we are supposed to resolve our LLRW dilemma by 1993?

Developing Our Radioactive Waste Policy

The history of the development of our national radioactive waste policy has been one of shortsightedness and neglect. While the failure to develop a coherent and workable national policy on radwaste disposal during the Manhattan Project may be understandable in light of wartime exigencies, the failure to do so in the 42 years since the end of the war is neither understandable nor acceptable.

In the immediate post-war period, almost all radwastes generated were from weapons research and production. Such wastes have always been exempt from regulation on the basis of national security. But the Atomic

Energy Act (AEA) of 1954 specifically authorized commercial uses of atomic energy without modifying the Atomic Energy Commission's (AEC's) existing regulatory authority over virtually all aspects of atomic energy. Since the AEA of 1954 was a direct result of intense lobbying to open the atomic energy field to commercial interests, it seems reasonable that the AEC should have anticipated that their existing hodgepodge waste disposal system would be overwhelmed by predictable increases in waste volumes.

In July 1955, a sanitary engineer from the AEC's Reactor Development Division addressed exactly this point:

> . . . One has only to consult the popular press to become acutely aware of the militant interest of the public in matters directly concerned with waste disposal and environmental sanitation. The increasing legal and legislative activity in this field is further evidence of the concern of the public and its acanthosis conclusion, it appears reasonable to state that the waste disposal problems associated with an expanding nuclear energy industry represent an important and integral part of any consideration of the industry's development. To date the problems have been adequately met, but when one views the future potentialities of the industry, it is obvious that safe, more efficient and economical ultimate disposal of radioactive wastes is one of the major challenges to the [atomic energy] industry as it progresses.[22]

A Regional System of Disposal Sites Was Never Developed

Recognizing that it made no sense from either an economic or safety point of view to transport LLRW any farther than necessary, the AEC began discussing the development of a regional network of disposal sites in the late 1940s. And though the Manhattan Project's disposal sites might loosely be characterized as a regional disposal network, the majority of commercial LLRW were disposed of at only two of the them, Oak Ridge, Tennessee and Idaho Falls, Idaho. When the agency decided to allow commercial disposal of LLRW, they continued to talk about the regional concept but failed to take any action to see that it became reality.

The Public's Fear of Radiation Has Influenced Our LLRW Policy

The public's fear of radiation is so widespread that it has been characterized a phobia. This concern has been exacerbated in the last few years by the extensive and, at times, sensational news coverage of the accidents at Three Mile Island and Chernobyl.

This fear is also exacerbated by Hollywood's use of radiation as a sort of an all-purpose bogeyman. For instance, in 1986, a movie called *The Manhattan Project* opened in theaters across the country. Aimed at the teen audience, the movie was aptly described by one reviewer this way, "The Manhattan Project [is] a low-key, high-tech, out-of-touch tale of *a teen who builds his own personal nuclear projectile for a science [fair] project.*"[23] (italics mine) (Actually, if I remember the plot correctly he builds an atomic bomb, not a projectile.) The young hero manages to enter (by defeating the alarms with Frisbees) a high-security laboratory which makes top-secret "super plutonium" for the military. He steals some of the plutonium and constructs an atomic bomb (in his garage) for a science fair project. The rest of the movie revolves around the young man's successful efforts to outwit the military, the NRC, and every other adult in the movie.

Now, if you want to attract the teen audience, a plot where a young hero makes authority figures look like idiots is a good way to do it. And I would never argue that movie scripts need be either reasonable or scientifically accurate. (I happen to enjoy science fiction.) What concerns me is that these inane, fear-provoking plots exacerbate the public's fear of radiation.

Not in My Backyard

LULUs

No one wants to live next door to a LLRW disposal facility. When a site for a disposal site is proposed, the cry NIMBY is inevitably heard. Unfortunately, as former EPA Administrator Douglas Costle once remarked, "The problem is that every place is someone's backyard." This observation is apparently still accurate in light of the public's adamant opposition to the siting of facilities such as chemical or radioactive waste disposal facilities, sanitary landfills, or municipal refuse incinerators. Even when local communities are willing, even anxious, to host a disposal facility, vehement opposition is often expressed by concerned citizens living tens or hundreds of miles away. Apparently, our backyards are growing all the time.

Waste disposal facilities, like airports, prisons, power plants, and FBI offices[24] are examples of what have been labeled "locally undesirable land uses" (LULUs). Siting LULUs has become a formidable social and legal challenge in today's litigious society. Citizens opposed to the facility can delay its siting for literally years by challenging every step of the process in the court system. The resulting delay may destroy the economic viability of a commercial venture and lead to cancellation of the project. Naturally, this is part of the opposition's strategy.

Part of the community's objection to a LULU is that the people in its immediate area have to bear the burden of the LULU though its benefits are widespread. For a LLRW compact, this means that the host state has to bear the burden of an entire geographic region.

People are simply afraid to live near these facilities. In addition to their fear of radiation, they are concerned about the quality of their lives. An individual who fears that his child will develop cancer or that he will lose a lifetime accumulation of equity in his home if a LULU is built in his neighborhood, is unlikely to be reassured by either government or private attempts to "educate" him scientifically. Any benefits that accrue to his community in terms of increased taxes and new jobs are not likely to offset these concerns.

Given that no one wants these undesirable, but necessary, facilities in their neighborhoods, where will we put them? We need municipal landfills, chemical waste handling facilities, and radioactive waste disposal sites; they are not options we can choose not to buy.

Siting LULUs away from population centers presents practical, economic, and sociological difficulties. As our population grows, we have fewer areas where the population density is low enough to site these facilities away from people. Years ago, when the majority of our population was concentrated in cities surrounded by rural areas of low population density, it was relatively easy to dispose of municipal wastes. Landfills were constructed outside of town in thinly populated areas where they upset no one. Today, suburbs are located where landfills and prisons would once have been located. And as a practical matter, certain types of LULUs, such as airports, need to be located near population centers to be efficient.

We Must Develop a Realistic Approach
to the Radwaste Problem

We are going to continue to generate radioactive wastes whether we like it or not, and we are going to have to dispose of them. This problem will not go away because people are afraid of radiation or because Congress passes a law. Nor is the radwaste problem, as some would like to portray it, a problem that would disappear if we stopped using nuclear power. It is true that the nuclear power industry generates a large percentage of our LLRW. But even if we had no nuclear power plants in this country, we would still generate over 500,000 cubic feet of LLRW per year from other nonmilitary sources. Where will we put this material? Are we willing to give up our consumer goods or lower our standard of living to minimize radwaste production? Does anyone want to go back to the days when heart attacks were diagnosed by hospitalizing people for days for observation rather than using radioimmunoassay techniques to determine cardiac enzyme levels? Can we do this in an era where dramatic increases in medical costs has led the federal government to set limits on the lengths of stay for hospitalized patients?

Should Radwaste Disposal Be a Federal Responsibility?

I argue that since society in general receives the social benefits of radwaste-producing activities, such as the development of new drugs, radiopharmaceutical diagnostic and treatment techniques, biomedical research, and nuclear power (until and if it is supplanted by another source), that disposal of radwastes should be a federal responsibility.

Why should Maryland be responsible for the disposal of LLRW from the National Institutes of Health (NIH), when, in 1984, NIH generated 43% of the state's LLRW volume. Surely the social benefits of the NIH's research are spread throughout the country. Indeed, they are the *National* Institutes of Health. Why does the federal government not take responsibility for this waste?

The federal government already has LLRW disposal sites on federal property located around the country. These sites are already in place, people around them have either accepted their existence, or do not know they are there. Why not use them instead of requiring each state to take responsibility for its waste?

Wastes That Are Below Regulatory Concern

Even if the government were to reassume responsibility for radwaste disposal, we need to make every effort to minimize waste production to conserve valuable natural resources such as disposal sites. In 1981, the NRC promulgated the so-called biomedical exclusion[25] to allow the disposal of very small quantities of certain isotopes "without regard to their radioactivity." This rule should be extended to other isotopes with low activity and short half-lives.

A controversial provision of the LLRWPAA of 1985 instructed the NRC to develop the standards, procedures, and technical capability to act on petitions from generators to exempt their waste streams from NRC regulations as "Below Regulatory Concern" (BRC) by July 1986. This is apparently proving to be a difficult task for the agency. In August 1986, they issued a policy statement and staff implementation plan that established procedures for groups of generators to petition the agency to declare their waste streams BRC. The guidance document also laid out the criteria that the agency staff would use in evaluating these petitions.

There are two problems with this plan. First, it requires generators to shoulder the burden of establishing that their wastes do not need to be regulated to protect public health, a job that the NRC and EPA should be doing. Second, while it appeals to certain classes of generators, it has disturbed a number of cities and counties that are already having difficulty siting disposal facilities and feel that this will make their task even more difficult. Several environmental groups have mounted campaigns against this rule.[26]

This plan was bound to arouse public opposition in view of the public's deep-seated fear of radiation. But I'm convinced that it would be better science and socially more acceptable to establish national standards on BRC wastes.

In May 1987, Texas adopted a BRC rule which will allow the disposal of radionuclides with half-lives less than 300 days in municipal landfills. This action was taken after an extensive study indicated that it was unlikely that any individual would receive a dose of radiation greater than one millirem, a dose level which even the EPA feels represents no significant risks to health.[27]

Through a combination of waste minimization efforts, the well-thought-out siting and operation of one or more national LLRW disposal sites, and the deregulation of truly low-level wastes that can be handled safely without extreme precautions, we could deal with this problem safely and efficiently. We should seriously consider taking these steps before we once again find ourselves on the edge of disaster.

REFERENCES

1. Helminski, E. L. "South Carolina Governor Tells States Barnwell Not Available," *The Radioactive Exchange* 6(9):2 (1987).
2. Helminski, E. L. *The Radioactive Exchange* 6(7):(1987).
3. Helminski, E. L. "Michigan Defers Decision On Host State Status Until August," *The Radioactive Exchange* 6(12):2–4 (1987).
4. "Michigan Chosen From Midwest Compact As Site For Low-level Waste Repository," *BNA Environment Reporter* 18(10):750–751 (1987).

5. Helminski, E. L. "California Adopts Four State Southwestern Compact, Includes Arizona," *The Radioactive Exchange* 6(11):1–2 (1987).
6. Helminski, E. L. "Texas Proceeds With LLRW Disposal Site Selection," *The Radioactive Exchange* 6(14):2 (1987).
7. Helminski, E. L. "Million Dollar Incentive Package, Strong Role for NY LLRW Site Host," The Radioactive Exchange 6(10):2–3 (1987).
8. "Recommendations for State Assistance to Localities Affected by The Siting of a Low-level Radioactive Waste Management Facility," Draft Report by The New York State Department of Environmental Conservation (April 1987).
9. Bord, R. J. *Opinions of Pennsylvanians on Policy Issues Related to Low-level Radioactive Waste Disposal* (University Park, PA: The Pennsylvania State University, 1985).
10. Bowerman, B. S., et al. "An Analysis of Low-level Wastes: Review of Hazardous Waste Regulations and Identification of Mixed Wastes," U.S. NRC Report NUREG/CR-4406 (1985).
11. "Joint EPA/NRC Guidance on the Definition and Identification of Commercial Mixed Low-level Radioactive and Hazardous Waste," USEPA Directive Number 9432.00–2 (1987).
12. "Joint Siting Guidelines for LLW Facilities Accepting MW Accepted by NRC and EPA," *Nuclear Waste News* 7(16):(1987) 95–96.
13. "Maxey Flat Respondents Enter Consent Order With EPA for Nuclear Site RI/FS," *Superfund Report Newswatch* 1(6):(1987) 1–2.
14. Peterson, C. "More Unwelcome Waste – This Time Radioactive," *Washington Post*:A3 (June 11, 1987).
15. "Radioactive Waste Production Declines," American Medical News:30, (May 15, 1987).
16. "Characteristics of Medically Related Low-level Radioactive Waste," U.S. Department of Energy Report, (1987).
17. Kelly, W. J., et al. "Barge Carrying Unwanted Garbage From Long Island Becomes Symbol for Larger Problem Of Solid Waste Disposal," *Environmental Reporter*:332–337 (May 15, 1987).
18. "Burning OK'd for Gar-Barge Contents," *Frederick, (MD) Post*:D8 (July 11, 1987).
19. "Trash Barge Solution Ordered," *Washington Post*:A9 (May 21, 1987)
20. "Garbage Barge at Final Stop: Where It Started," *The Washington Post*:A3 (July 11, 1987).
21. "Fear Of Disease Problem For Barge," *Frederick (MD) Post*, (July 13, 1987).
22. Lieberman, J.A. "Disposal of Radioactive Wastes" *Civil Engineering* 25:44–47 (1955).
23. Kempley, R. "The Manhattan Project Builds A Stinkbomb," *The Washington Post*, (June 13, 1986).
24. In Washington, DC a local hospital and a group of neighborhood activists have banded together to try to block the establishment of an FBI counterintelligence office in their neighborhood. Their objections allegedly revolve

around the possibility of FBI radio equipment interfering with cardiac monitors in the hospital and the possibility that the office might become a terrorist target. It may help to put these concerns in perspective to learn that the Drug Enforcement Administration, which is probably a far more likely candidate for a terrorist attack, already has an office in the same building! "Fauntroy Bill Would Block FBI on M Street Site," *The Washington Post*:B3 (August 4, 1987).

25. The biomedical exclusion allows the disposal of certain organic solvents or animal carcasses contaminated with low levels of tritium and carbon-14 "without regard to their radioactivity." Rosalyn Yalow and other nuclear scientists have argued persuasively that this rule should be extended to other isotopes, especially those with short half-lives.

26. "Below Regulatory Concern" Criteria Proves Difficult Job For NRC, EPA" *Nuclear Waste News* 7(20):127–128 (1987).

27. Helminski, E. L. "Texas Adopts BRC Standards For LLRW Allowing Landfill Disposal," *The Radioactive Exchange* 6(9):3 (1987).

Glossary

activity (radioactivity)   the number of radioactive transformations that take place in a given period of time. Measured in becquerels or curies, where one curie = 37 billion disintegrations per second = 37 billion becquerels.

agreement states   states that have authority from the Nuclear Regulatory Commission to promulgate and enforce their own radiation regulatory programs. These regulations are required to be at least as stringent, and maybe more so, as the federal regulations.

ALARA (as low as readily achievable)   the concept that all radiation exposures should be kept to a minimum, while taking into account the state of technology, the economics of improvements in relation to public health and safety, other societal and socioeconomic considerations, and the relation of the above to the utilization of atomic energy in the public interest.

alpha particle   a positively charged particle emitted from certain nuclei (typically, transuranic elements) during radioactive decay. An alpha particle is identical to the nucleus of a helium atom, containing two protons and two neutrons and having an atomic mass of four atomic mass units. Alpha particles are only capable of traveling a few inches in air and are stopped a sheet of paper or intact skin.

background radiation   radiation we are exposed to in the environment. Sources include cosmic rays and naturally occurring radioactive elements outside of or in our bodies. For example, our bones inevitably contain a small percentage of a radioactive isotope of potassium (K-40) and nonradioactive K-39.

becquerel (Bq)   the SI system unit of radioactivity. One becquerel is equal to 2.7×10^{-11} curies.

beta particle   a negatively charged particle emitted by a nucleus during radioactive decay. It has mass and charge equal to that of an electron. Depending upon the energy with which they are emitted, they can be stopped by a sheet of plastic or a few inches of water.

curie (Ci)   a historical unit of radioactivity. One Ci is equal to the number of disintegrations given off by one gram of radium in one second (37 billion disintegrations per second). The curie has been replaced by the SI unit becquerel (Bq), where one curie = 37 billion becquerels.

decay (radioactive)   the decrease, due to spontaneous disintegration, in the amount of radioactive material present as time passes. As it decays, the radioisotope changes to lighter, more stable elements. Eventually, the process ends in the formation of a nonradioactive element or isotope.

decay series   the specific steps undergone by a radioisotope as it decays to a stable structure. For instance, the first step in the radioactive decay of radium-226 is the emission of an alpha particle, resulting in the formation of an atom of the gaseous element radon-222. Eventually, after many years the original radium atom becomes an atom of nonradioactive lead-210.

dry active waste   radioactively contaminated solid wastes such as gloves, protective suits, and paper towels.

electron   a negatively charged subatomic particle that is found outside an atom's nucleus. Its mass is 1/1837 that of a proton and can thus be considered negligible for most applications.

fission   the splitting of an atomic nucleus into two or more parts called fission fragments. In the process, a tremendous amount of energy is released as a tiny amount of mass is converted into energy.

fission products (fragments)   the lighter weight products of a fission reaction.

gamma ray   a high energy emission given off by some atomic nuclei during radioactive disintegration. It has properties similar to those of the X-ray. Gamma rays are capable of penetrating solid objects, though they can be shielded by adequate thicknesses of lead.

gray (Gy)   the SI unit of absorbed radiation dose. One gray is equal to one joule per kilogram, or 100 rad.

half-life ($t_{1/2}$)   the amount of time it takes for one half of any amount of a radioactive isotope to decay. Since the rate of disintegration of any isotope is a constant, the remaining amount, at any time, can be calculated using the exponential decay function. Half-lives may be extremely short, on the order of fractions of a second, or very long, lasting billions of years. Low-level radioactive wastes contain both long-and short-half-lived isotopes.

ion   an atom that has acquired either a positive or negative electrical charge by losing or gaining one or more electrons.

ion exchange resin   a medium capable of removing ions from a solution and replacing them with another ionic species. Ion exchange resins are used in many home water softeners, where they remove cations such as magnesium and copper and replace them with sodium ions. In nuclear power plants, they remove radioactive ions from cooling water.

ionizing radiation   any radiation capable of displacing electrons from atoms or molecules to form ions.

isotope   atoms of a given element (i.e., they have identical atomic numbers) which differ in atomic mass. For example, hydrogen exists naturally in two isotopes, ^1H and ^2H (deuterium). Deuterium can, in turn, be bombarded with neutrons to form the radioactive isotope ^3H (tritium).

neutron   a subatomic particle found in the nucleus of an atom. It carries no electrical charge.

proton   a positively charged subatomic particle found in the nucleus of an atom. The number of protons in an element's nucleus is equal to its atomic number.

rad (radiation absorbed dose)   the historic unit of absorbed radiation. The rad has been replaced by the SI system unit gray (Gy), with one Gy = 100 rad.

radiation   the emission of a particle, energy, or both, during radioactive decay.

radioactive decay (radioactivity)   the process by which an unstable atomic nucleus loses energy by emitting a particle and/or energy in order to increase its stability. In the process, it is transmuted to a different radioisotope. For instance, the first step in radioactive decay of the radium-226 atom is the emission of an alpha particle, resulting in its transmutation to radon-222. Any given radioisotope will disintegrate at a specific rate—a specific number of transformations per unit of time (usually per second). This is known as the isotope's specific activity.

radioisotope   an unstable isotope that spontaneously emits particles and/or energy in order to reach a stable state.

rem (roentgen equivalent man)   the historic unit of radiation dose for humans. The rem has been replaced with the SI unit sievert (Sv) with one sievert = 100 rem.

roentgen (R)   the historic unit of radiation exposure.

scintillation cocktail   any of a number of chemical solvents that contain small amounts of radioluminescent materials. When combined with a radioisotope (e.g., P-32), the solution will first absorb energy and then release it as fluorescence. The intensity of the emission varies between radioisotopes and is proportional to the amount of radioisotope present. The process, called scintillation counting, is widely used in today's biomedical research.

sievert (Sv)   The SI unit for a radiation dose equivalent. One Sv = 100 rem.

transuranics (transuranic elements)   elements with atomic numbers greater than that of uranium, i.e., greater than 92.

Abbreviations

AEC	Atomic Energy Commission
ALARA	as low as readily achievable
BEIR	biological effects of ionizing radiation
Bq	becquerel (see glossary)
BRC	below regulatory concern
BWR	boiling water reactor
CERCLA	The Comprehensive Environmental Response, Compensation and Liability Act of 1980 (Superfund)
CFR	Code of Federal Regulations
Ci	curie
CPSC	Consumer Product Safety Commission
DAW	dry active wastes
DEC	Department of Environmental Conservation
DER	Department of Environmental Restoration
DHHS	Department of Health and Human Services
DOE	Department of Energy
DOT	Department of Transportation
EIS	environmental impact statement
ELM	Environmental Lobby of Massachusetts
EPA	Environmental Protection Agency
EPRI	The Electric Power Research Institute
FDA	Food and Drug Administration
FEIS	final environmental impact statement
FIFRA	Federal Insecticide, Fungicide and Rodenticide Act
FR	Federal Register
GAO	General Accounting Office
GCD	greater confined disposal
Gy	gray (see glossary)
HHS	Department of Health and Human Services (formerly

	HEW)
HIC	high-integrity containers (for shipment of HLRW)
HLRW	high-level radioactive wastes
IAEA	International Atomic Energy Agency
ICRP	International Commission on Radiological Protection
IDNS	Illinois Department of Nuclear Safety
LET	linear energy transfer
LLRW	low-level radioactive waste
LLRWPA	Low-Level Radioactive Waste Policy Act of 1980
LLRWPAA	Low-Level Radioactive Waste Policy Amendments Act of 1985
LULU	locally undesirable land use
LWV	League of Women Voters
MED	Manhattan Energy District (pseudonym for the Manhattan Project)
MEPA	Massachusetts Environmental Policy Act
mrem	millirem
NAS	National Academy of Sciences
NCI	National Cancer Institute
NCRP	National Council on Radiation Protection and Measurements
NDA	New Drug Application
NEPA	National Environmental Policy Act
NFS	Nuclear Fuel Services, Inc.
NIEHS	National Institute of Environmental Health Science
NIMBY	not in my backyard
NRC	Nuclear Regulatory Commission
NSF	National Science Foundation
NYDEC	New York Department of Environmental Conservation
NYSERDA	New York State Energy Research and Development Authority
NYSGS	New York State Geological Survey
OMB	Office of Management and Budget
ORGDP	Oak Ridge Gaseous Diffusion Plant
ORNL	Oak Ridge National Laboratory
OSHA	Occupational Safety and Health Administration

OSTP	Office of Science and Technology Policy
PL	public law
PPB	parts per billion
PPM	parts per million
PWR	pressurized water reactor
PWS	public water supply
RBE	relative biological effectiveness
RCRA	Resource Conservation and Recovery Act
SAB	Science Advisory Board (EPA)
SAT	Scholastic Aptitude Test
SDWA	Safe Drinking Water Act
SI	Systeme International d'Unites (International System of Units)
SMR	standardized mortality ratio
Superfund	The Comprehensive Environmental Response, Compensation and Liability Act of 1980
Sv	sievert (see glossary)
TMI	Three Mile Island
TRU	transuranic (elements or wastes having an atomic number greater than 92)
TSCA	Toxic Substances Control Act
TSSC	EPA's Toxic Substances Strategy Committee
USC	United States Code
USGS	United States Geological Survey
VLLW	very low-level radioactive waste
VR	volume reduction

Index